PLANES and MODEL KITS

FOCKE WULF FW 190A/F

ASSEMBLY - TRANSFORMATION - PAINTING - WEATHERING

Pedro ANDRADA, Aurelio GIMENO, Alfonso MARTINEZ BERLANA and Eduardo SOLER

CONTENTS

THE BIRTH OF A LEGEND Eduardo Soler, Aurelio Gimeno and Alfonso Martinez Berlana	4
THE FOCKE-WULF IN KIT FORM Eduardo Soler	6
A GUIDE TO FW 190 VERSIONS Aurelio Gimeno, Eduardo Soler	8
COLOUR SCHEMES OF THE FW 190 Aurelio Gimeno	11
THE FW 190 V-1 THE ADVENTURE BEGINS Aurelio Gimeno	13
THE FW 190A-4 CHECKMATE Eduardo Soler	24
THE FW 190A-5, AND THE SUPERIORITY OF THE BMW ENGINE ON THE RUSSIAN FRONT Alfonso Martinez Berlana	35
THE FW 190R-8, THE «PANZERBLITZ» IN ACTION Aurelio Gimeno	48
A CAMOUFLAGE SCHEME Aurelio Gimeno	68
THE DEFENCE OF THE REICH BANDS Eduardo Soler	71
COWLING COLOURS Eduardo Soler	72
PROFILES AND ILLUSTRATIONS Eduardo Soler and Aurelio Gimeno	74
BIBLIOGRAPHY - THANKS	82

HISTOIRE & COLLECTIONS

FOCKE WULF Fw 190
THE BIRTH OF A LEGEND

An Fw 190 S-8, the twin seater version of the Fw 190 A-8/U-8 from a fighter flying school (Jagdflierschule) end of 1944. The camouflage is RLM 74/75/76 with a green - black spinner in RLM 70.

The origins of the Fw 190 go back to 1937 at the request of the (RLM) German Air ministry for a second-generation monoplane to replace the BF109 when the time came.

The request was made directly to the Focke Wulf aircraft factory and responsibility for the project was placed in the hands of the prestigious engineer Kurt Tank. The choice of engine was either to be the in-line DB 601 or the radial BMW 139. BMW was in fact chosen as the production was more advanced, BMW were also able to handle a larger production capacity.

The V- 1 prototype's maiden flight took place on June 1 1939 at Bremen, piloted by Hans Sanders. The prototype was equipped with a high and wide tracked landing gear, an all round vision canopy, a well-designed engine cowling and an elegant ducted propeller spinner. The plane's overall aspect was a radical departure from its predecessor and direct rival, the BF 109. However, in spite of its good manoeuvrability and various other qualities, it suffered from overheating caused by the ducted spinner. This problem, revealed in the first test flights, made the engineers opt for a more classic configuration and the V1 changed shape with a smaller spinner and a duct with a larger opening.

The first flights in operational conditions took place at Le Bourget with the 11./JG26. Several problems became apparent, the engine-overheating problem resurfaced, caused by a failure in the propeller pitch system. A commission sent by the Air Ministry to examine the problems recommended that the Fw 190 programme should be abandoned. Someone even commented that *"flying the 190 was a bit like having one's feet in an oven."*

These recommendations were not followed up however, and it took no less than fifty modifications to resolve the Fw 190's various teething problems before it was declared fit for service.

Even though the BMW 801C engine improved its flying capabilities, the first series of planes were still heavily criticised for the lack of firepower. They were, indeed, armed with only four synchronised MG17 7.92 calibre machine guns and this was deemed insufficient by the pilots more used to the mixed cannon and machine gun configuration of the BF 109.

The armament was thus upgraded by the addition of two 20 mm cannon in the wing roots. The planes became operational over the Channel and were met with surprise by the British who initially thought they were captured French Curtiss H-75s. It soon became evident that their MKV Spitfires had lost their supremacy over the Channel, a situation that was to last for eight months until the arrival of the improved versions of the Spitfire. However, even though the radial BMW 801cc (1560 horse power) engine helped the first Fw 190A-1 to a very respectable top speed of 624 kph, it became clear that it was less suited to high altitude flying than the in-line Daimler Benz and this meant that it never really replaced the Messerschmitt that was powered by the Daimler Benz. The versatility of its modular design, on the other hand, allowed the development of many versions, variants and sub-variants particularly suited to the needs of the Luftwaffe, whose ground attack aircraft were particularly vulnerable.

These evolutions corresponded in many cases to the roles undertaken by the BF 109 and Kurt Tank's plane excelled notably in a ground attack role and against Allied four-engined bomber formations. In both situations the radial engine ensured not only a better protection to the pilot but was also less vulnerable to enemy fire being air-cooled instead of liquid-cooled by glycol. Even though fewer 190s were manufactured than the Bf 109 (a little more than 20,000 compared to 30,480) and that certain highly decorated pilots preferred the latter, other famous aces such as Otto Kittel (267 kills), Erich Rudorfer (222 kills), Heinz Bar (220) or Walter Nowotny (258) declared that the Fw 190 was superior in many ways to the Bf 109 and gave it their seal of approval.

Another ace, Gunther Rall, third highest scoring pilot of the Second World War with his 275 kills and used to the Bf 109, considered the Fw 190 superior due to its high speed and excellent turning capabilities. Adolf Galland, the Luftwaffe's youngest general even said that its production ought to take precedence over that of the BF 109.

We can add to this its robustness, easy maintenance and a simple, yet modern design, not forgetting its capacity to carry out a multitude of missions (the Fw 190 was easily the most versatile plane in the Luftwaffe). This enabled it to be fitted at the most forward of airfields with bombs, drop tanks, rockets, cannon of different calibres and even torpedoes, rendering unnecessary a return to the factory.

These sub variants were recognisable throughout the war by the attribution of the letter R (for Rüstsatz, roughly meaning field upgrade kits). Thus equipped, planes were able to carry out specific missions, like the Fw 190 A-5/R6 armed with 21 cm W. GR 21 rockets derived from mortar shells that caused havoc amongst B17 Flying Fortress formations. This type of kit, previously recognised by the letter U (Umrüst Bausätze, factory upgrade kits) were found on the Fw 190 A-5/V14, armed with a 950 kg torpedo and flown by Lt Helmut Viedebannt of the SKG 10 over the Channel.

In action

The presence of the Fw 190 over the Channel and its obvious qualities worried the top brass of the RAF to such an extent that it was the direct cause of the Spitfire upgrades and modifications that led to the excellent Mark IX that saw out the end of the war (a rare example of longevity for a version that first appeared in mid 1942). Spitfires were also urgently dispatched to North Africa in early 1943 to combat the Fw 190. Apart from its use on the Western Front, it also played an essential role on the Eastern Front, particularly in Russia where it came into its own not only as a fighter but also more importantly as an assault and ground attack plane, replacing the Stuka and Henschel Hs129 in extreme weather conditions where it was less vulnerable and, more importantly, more capable of self defence than the latter two aircraft.

It also confronted, with different armament configurations, and various protective and detection equipment, the heavy bombers of the USAAF that flew daylight missions over Germany. The Fw 190 was also sent against the night bombers of the RAF and the night time raids against civilian targets. A 'wild boar' tactic was employed, the idea being of a sow charging madly in defence of her young, which was supposed to sum up the attitud of the Luftwaffe pilots towards their mission.

Finally we must not forget the specific bombing missions of the Fw 190. The undercarriage was strengthened and the plane was able to carry bomb loads of 1,000kg, 1,600kg and occasionally 1,800kg, a weight that would have been almost out of the question on a late 1930s monoplane.

In 1942 its modern design made it possible, in order to improve its performance at altitude, to install an in-line liquid cooled engine that was to equip the Fw 190D, eventually making the nose longer.

The Ta 152 H, the last in a long line, included important modifications, which enabled it to fly at high altitude at the speed of 760 kph.

This gave it superiority over most of its rivals including the formidable American Mustangs.

Rechlin, the Luftwaffe test flight centre, summer 1939. Next to the D- OPZE Fw 190V-1 prototype we can see Flugkapitän Lucht and General Ernest Udet, head of technical development. Udet said that *"If it wasn't a fighter plane it would be magnificent"*. Next to him is engineer Karl Francke, one of the Fw 190s test pilots with Hans Sanders who took the Fw 190 up for the first time. *(RR)*

The V-1 Fw 190 prototype was heavily modified from January 25th 1940, essentially in its cooling system with the replacement of the aerodynamic prop spinner by a more conventional but more efficient spinner that became standard. It was given the code FO°LY, crosses were added and a conventional swastika added to the tail plane as well as a small white 01 on the rudder. *(RR)*

THE FOCKE WULF FW 190 MODEL KITS

The different types of radial engined Fw 190s have been fairly well covered by kit manufacturers over the years, from the once extremely rare 1/24 scale to the tiny 1/144. No one kit really stands out from the others. Instead there are several high quality products to which we can add the recent *Hasegawa* made 1/32 scale series of the Fw 190A/F.

Listing all the kits would be a long and tedious process, especially as some are one-offs or out of stock in most shops. We have instead made a selection based on the quality and availability of the kits, some of which are no longer made, as well as on the number that were made. This was done for each main scale on the market.

Credit...

Let us begin with the 1/24 scale and the *Airfix* model that appeared at the beginning of the 1980s, later made by Heller, and what was really the great British manufacturer's swan song in this scale. This relatively detailed kit that can be displayed with its cowling and ammunition bays open and engine visible is still unique as we write these lines and the only way to tackle the Fw 190 in this scale.

Moving on to the 1/32 scale we find the first *Hasegawa* kit also made by the English manufacturer *Frog* then by *Revell*. This kit made its mark in the seventies, offering varying possibilities using as its base the Fw 190A-5 (with a fuselage 15 cm longer compared to previous versions) an option that allowed to go as far as the Fw 190A-8 for example, by making the appropriate modifications without any major structural changes.

It was, however, let down by some considerable faults. Ignoring the weakness of detail and the structural lines that literally stood out, the most important fault was in the undercarriage wells, which were particularly inaccurate. One of its best qualities, however, was the BMW 801 engine that made it easier to display in various scenarios.

In the 1/48 scale we have to mention the very old kit made by America's *Monogram* in the late 1960s. It was a pleasant surprise to modellers as, like with the *Hasegawa* kit in 1/32, it was possible to build different versions thanks to its optional parts and plentiful decals. Today it seems old fashioned, at least by modern standards, as certain details such as the interior of the undercarriage wells are lacking, not to mention the outmoded protruding rivets. Even so it made its mark on the world of model kits. Another level was later reached by the Japanese 1/48 scale *Otaki* kit, reproduced later by *Arii* as well as *Matchbox* and *Airfix* in England. This was the A-8 version and for the first time in this scale, panel lines were etched. It needed a lot of improvements and the addition of certain details in the undercarriage wells but it was all the same a good starting point for making a very acceptable model.

... where credit is due

Lastly, *Fujimi* launched, around the same time, a fairly rudimentary version of the A-8 with etched structural lines but shapes that didn't really conform to reality. Still in the 1/48th scale, the arrival of the *Trimaster* 190 A-8 in the late 1980s marked a turning point in the world of aircraft modelling. It was in fact a version derived from the previously released Fw 190D and included photoetched and white metal parts, a first for a mass-produced kit.

There were a lot of decals, great quality moulding and a near perfect respect to detail, especially the undercarriage wells. We should note, however, that the original model presented several assembly problems with certain parts, thus limiting this kit, considered as historic and exceptional, to the advanced modeller. This mould was later used by *Dragon*, *Italeri* and *Revell* who offered it in almost all versions except those of the short fuselage Fw 190 A-1 to A-4. After this, top marks go to Tamiya who cleverly produced for the first time the A-3 version with the unrivalled expertise for which they are known. The *'Brand with the star'* then focused on the radial engined F-8, using this as a base and with several easy modifications you could create all the long fuselage versions that started with the Fw 190 A-5.

We must not forget *Trimaster* again, who marketed the Ta 152 (a distant cousin of the Fw 190 A) a variant not dealt with here. This was remade by Italeri but without the refinements of the white metal and photoetched parts, which made it considerably cheaper and therefore accessible to a wider public.

Today, this kit is still the best in this scale and is often seen in shows.

Still popular

The much-liked 1/72 scale was also well served by a multitude of kits. In 1959, *Frog* was the first manufacturer to release a commercial kit of the Fw 190 (this does not take into account the wartime identification models). Lacking undercarriage wells it only vaguely resembled the German fighter and was fairly rudimentary and uninteresting for keen modellers.

Next to come was a *Revell* kit of a badly defined version, the Fw 190A/F by Heller with two canopies, flat and bulged and the *Hasegawa* kit with its long fuselage that is a far cry from its more recent productions.

In 1979, *Matchbox* launched an A-3/A-4 cast in injection moulded garish blue and green plastic. At the time it was the most accurate model in this scale in spite of its deep etched structural lines that overcrowded its surface and extremely limited attention to detail.

It was *Airfix*'s turn in the late seventies to produce an F-8 with a reasonably realistic outline despite its embossed lines characteristic of the time. The following decade *Airfix* marketed a considerably modified and improved A-8/F-8. *Italeri*'s Fw 190 A-8 was this version's standard-bearer in the early 1980s in spite of some inadequacies, mostly concerning detail. Towards the mid 80s *Hasegawa* launched a new edition of the Fw 190A-5, heralding the arrival of a new generation of quality kits.

This model, as is standard practice of the Japanese firm, offered several versions, some quite rare, with drop tanks, Bt400 missiles and a range of ace decals such as H. Bar, P. Priller, A. Galland or H. Graff.

We also have to mention the excellent kit made by *Monogram- Promodeler/Revell* that allows the building of versions from the A-6 to the F-8 with very up-to-date etching and attractive detail. Thanks to the conversion kits you can transform the kit into different variants armed with torpedoes and glide bombs.

Let's go back again now to *Tamiya*, who pantographed its Fw 190 A-3 in the 1/48 scale resulting in a superb model which, with a minimum amount of work can depict the very early variants.

We can next mention the Korean make *Academy* and the Czech *Smer* who marketed kits of the Fw 190 that had some inaccuracies compared to the previously mentioned makes. We finish now with some lesser known makes like *Sword* with its A-1 or MPM that produced the V-1 prototype. These are unique kits in our opinion, not so much by their quality but because they are the only two injection moulded examples of these two rare versions, without forgetting the ' Jabo'version of the Fw190 A-4 by Admiral that contains some resin parts.

Overdone?

So at the end of this rapid review the Focke Wulf 190 can be considered to be one of the best represented aircraft by both kit manufacturers and book publishers. All this proves how justifiably important and popular it is. At the time of writing we decided to opt for the 1/48 scale that allows, in our opinion the best compromise between attention to detail and general dimension. The choice of model, colour scheme and decals were more difficult as each plane needed to be both a good representation of the Fw 190's long career, different enough to offer an original composition and at the same time requiring different techniques, from the complete scratchbuilt to the usual minimum attention to detail.

We logically begin with the first prototype, a scratchbuild using a plaster block. We move on successively and in detail from the Fw 190 A-4 to the A-5 and the F-8, three models from the Tamiya range. Each one of these has been made as detailed as possible, choosing the best suited and most accurate accessories from the huge range available.

THE BEST?

Trimaster, a brand that appeared in the late eighties, made the best Fw 190 in 1/48 to hit the shops. It started the trend of a new generation of models of exceptional fineness, a larger and higher quality decal sheet and photoetched parts. It was also at the origin of white metal parts, no longer solely resin, as well as other various additions such as metallic tubes or steel wires.

Suffering from marketing problems, Trimaster sold the moulds to manufacturers such as *Dragon* and *Revell*. Its interesting Fw 190 versions remain famous such as the A-8/R 11 night fighter.

The guide on pages 8 to 10 allows the identification of various Fw 190 versions and their main modifications.

Other variants existed within sub-variants 'U' and 'R', resulting in a multitude of differences with one same variant. We must also take note of certain distinctions with the armament and canopy, the A-8 and F-8 for example that were equipped with either a flat or bulged canopy.

The Focke Wulf Fw 190 A-4, a realistic transformation from the Tamiya's kit.

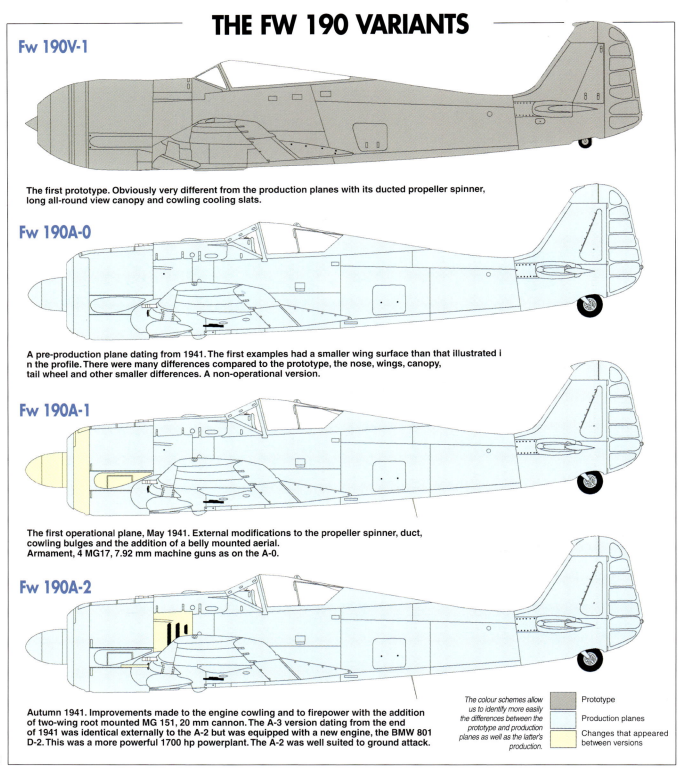

THE FW 190 VARIANTS

Fw 190V-1

The first prototype. Obviously very different from the production planes with its ducted propeller spinner, long all-round view canopy and cowling cooling slats.

Fw 190A-0

A pre-production plane dating from 1941. The first examples had a smaller wing surface than that illustrated in the profile. There were many differences compared to the prototype, the nose, wings, canopy, tail wheel and other smaller differences. A non-operational version.

Fw 190A-1

The first operational plane, May 1941. External modifications to the propeller spinner, duct, cowling bulges and the addition of a belly mounted aerial.
Armament, 4 MG17, 7.92 mm machine guns as on the A-0.

Fw 190A-2

Autumn 1941. Improvements made to the engine cowling and to firepower with the addition of two-wing root mounted MG 151, 20 mm cannon. The A-3 version dating from the end of 1941 was identical externally to the A-2 but was equipped with a new engine, the BMW 801 D-2. This was a more powerful 1700 hp powerplant. The A-2 was well suited to ground attack.

The colour schemes allow us to identify more easily the differences between the prototype and production planes as well as the latter's production.

- Prototype
- Production planes
- Changes that appeared between versions

Fw 190A-4

July 1942. This version was recognizable by an aerial mast mounted on the tail fin and a change in engine to the BMW 801 C-2 replacing the D model that suffered from a slow production rate. The last versions were equipped with mobile engine cooling slats as seen on the following A-5.

Fw 190A-5 – Fw190S-5

The Fw 190A-5 appeared at the beginning of 1943 along with its two-seater variant the S-5. The A-5 and the following versions were 15.25 cm longer and modifications were made to the flaps, rudder, fuselage maintenance hatch and the main panels under the tail plane.

Fw 190A-6

The Fw 190A-6 began production in May 1943 with a new armament (replacement of the wing mounted machine guns by four MG 151/20, 20 mm cannon.) We can also see the FuG 16ZE loop aerial.

Fw 190A-7

Began production in December 1943. The main external change is a bulged cowling that now housed two MG 131, 13 mm heavy machine guns (replacing the MG 17). The undercarriage doors were also slightly modified. The A-7 also had a fuel tank attached to an ETC 501 rack.

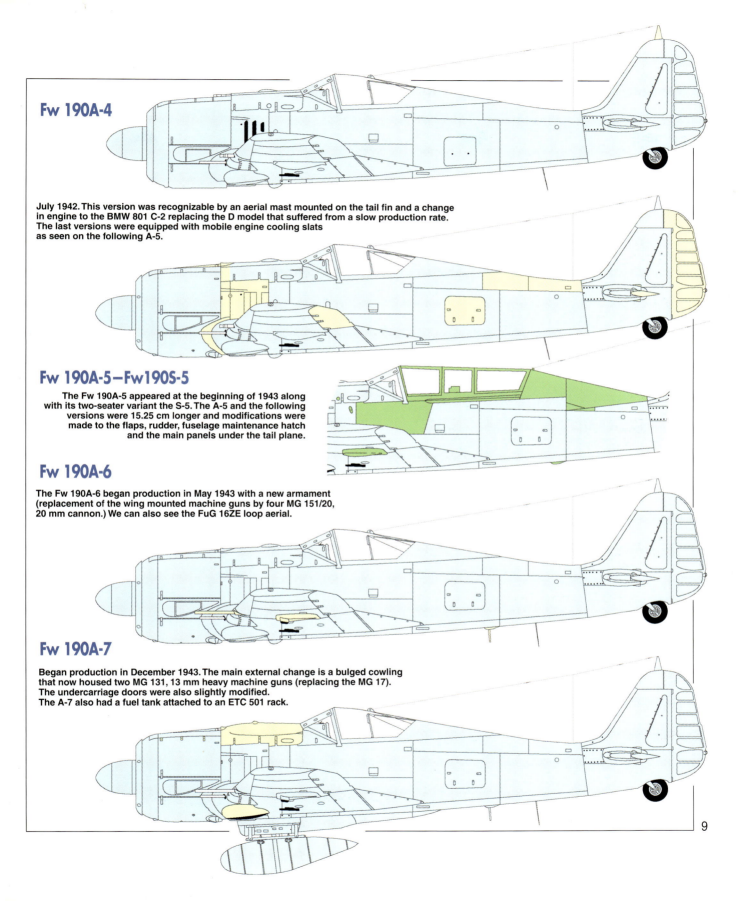

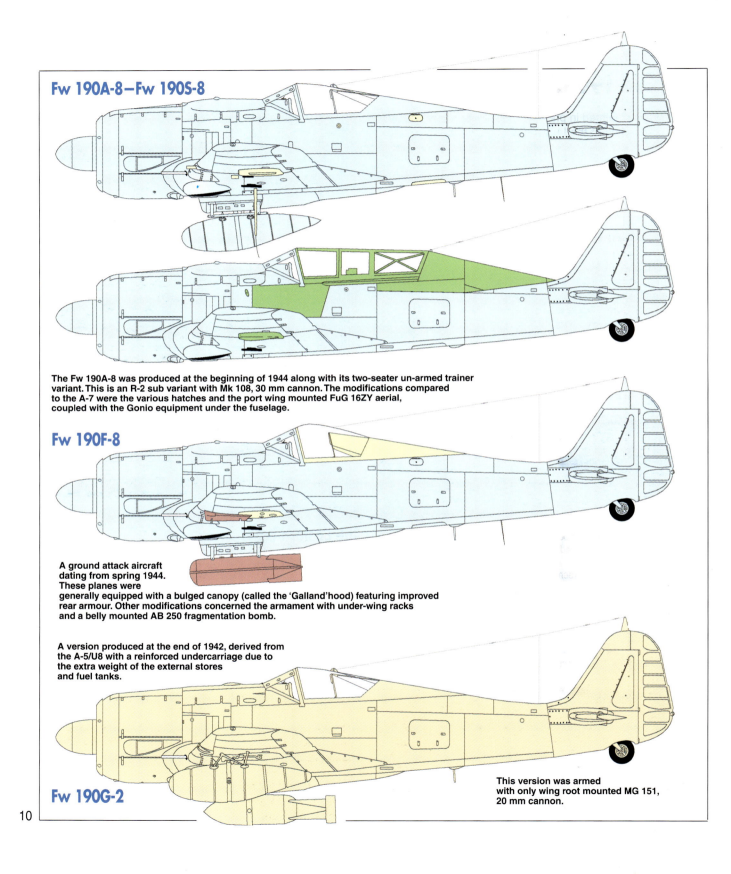

Fw 190A-8 — Fw 190S-8

The Fw 190A-8 was produced at the beginning of 1944 along with its two-seater un-armed trainer variant. This is an R-2 sub variant with Mk 108, 30 mm cannon. The modifications compared to the A-7 were the various hatches and the port wing mounted FuG 16ZY aerial, coupled with the Gonio equipment under the fuselage.

Fw 190F-8

A ground attack aircraft dating from spring 1944. These planes were generally equipped with a bulged canopy (called the 'Galland'hood) featuring improved rear armour. Other modifications concerned the armament with under-wing racks and a belly mounted AB 250 fragmentation bomb.

A version produced at the end of 1942, derived from the A-5/U8 with a reinforced undercarriage due to the extra weight of the external stores and fuel tanks.

Fw 190G-2

This version was armed with only wing root mounted MG 151, 20 mm cannon.

THE FW 190 COLOUR SCHEMES

The 190 was painted in practically all the standard and non standard schemes of the Luftwaffe, including the first colour schemes RLM 65, 70 and 71 applied to certain prototypes (see the V-1 kit in this book) or on planes on the Eastern Front where weather conditions meant a certain easing up in the following of the official RLM directives.

From the beginning, on the Eastern Front and up to 1944

The first examples of the A-1 version appeared in French skies in September 1941 and bore the official RLM 76, 75 and 75 colour scheme of the time with varying amounts of mottling on the fuselage and tail fin.

This type of camouflage was common to most German fighters of the period and was usually painted in RLM 02/70/74 but in the case of the Fw 190, only one or two colours were used. Sometimes the mottling could be totally absent, very light or

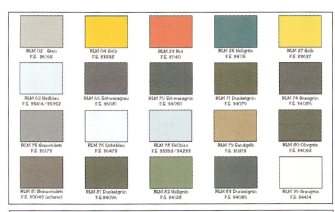

OFFICIAL RLM (GERMAN AIR MINISTRY) CHART

RLM 02	Grau	F.S. 36165	RLM 80	Olivgrün	F.S. 34052
RLM 04	Gelb	F.S. 33538	RLM 81	Braunviolett 1	F.S. 30045
RLM 23	Rot	F.S. 31140			(Lightened)
RLM 25	Hellgrün	F.S. 34115	RLM 81	Dunkelgrün	F.S. 34096
RLM 27	Gelb	F.S. 33637	RLM 82	Hellgrün	F.S. 34128
RLM 65	Hellblau	F.S. 35414	RLM 83	Dunkelgrün	F.S. 34086
		/35532	RLM 99	Graugrün	F.S. 34414
RLM 66	Schwarzgrau	F.S. 36081			
RLM 70	Schwarzgrün	F.S. 34050			
RLM 71	Dunkelgrün	F.S. 34079			
RLM 74	Graugrün	F.S. 34086			
RLM 75	Grauviolett	F.S. 36173			
RLM 76	Lichtblau	F.S. 36473			
RLM 78	Hellblau	F.S. 35352			
		/34233			
RLM 79	Sandgleb	F.S. 30215			

Note. Official colours used for the kits and profiles. Modern printing techniques cannot render a perfect copy of the colours and these are only an indication. Modellers are therefore advised to directly consult the Federal Standard colour chart.

OFFICIAL FW 190 PAINT SCHEMES

RLM 02 (grey, green) prototypes, sometimes applied to the exterior and interior areas such as the cockpit until November 1941.
RLM 04 (yellow) tactical markings, fuselage bands, wing tips, engine underside, rudder etc. Badges and numbers
RLM 23 (red) tactical markings, fuselage bands engine underside, rudder etc. badges and numbers
RLM 25 (light green) badges, numbers, bands and when needed, applied to certain camouflage
RLM 27 (yellow, lighter and greener than RLM 04), tactical markings, bands, wingtips, engine underside, rudder etc.
RLM 65 (light blue) lower wing surfaces, combined with RLM 70 and 71 greens on prototypes and certain planes in Russia.
RLM 66 (very dark grey) cockpit interior from November 1941.
RLM 70 (very dark green) upper wing surfaces in combination with RLM 71
RLM 71 (dark green) upper wing surfaces in combination with RLM 70 on prototypes and some aircraft in Russia.
RLM 74 (dark grey) upper wing surfaces in combination with RLM 75 (the most frequently used camouflage on all fronts).

RLM 75 (grey violet) upper wing surfaces combined with RLM 74 and occasionally with RLM 82 from 1944.
RLM 76 (light grey blue) upper wing surfaces used in various combinations (RLM 74 and 75 or RLM 81 and 82).
RLM 78 (bright sky blue) lower wing surfaces, combined with RLM 79 and 80 in the Mediterranean theatre.
RLM 79 (sand) upper wing surfaces, used alone or with RLM 80 in the Mediterranean theatre.
RLM 80 (olive green) upper wing surfaces, often used for a mottled effect on RLM 79. Mostly used in the Mediterranean theatre.
RLM 81 (brown violet or green brown depending on the contractor) upper wing surfaces combined with RLM 82 from 1944.
RLM 82 (mid green) upper wing surfaces in combination RLM 81 or RLM 75 from 1944.
RLM 83 (dark green) upper wing surfaces in combination with RLM 81 from 1944.
RLM 99 (very pale green) upper wing surfaces. According to certain authors, this shade never existed. But it could be a variation of RLM 76 like RLM 84.

almost transparent. This camouflage remained in use until June 30th 1944, except for a few variants destined for night fighting or other very specific missions.

The Eastern Front

The Fw 190 was widely used in Russia due to its solid design that could withstand considerable punishment and easy maintenance. This particular front led to all sorts of camouflage schemes as we mentioned above.

The early basic colour scheme was composed of RLM 76, 75 and 74 but the harsh winter conditions imposed certain practical changes. A white washable paint was applied over the original camouflage and helped the plane to merge into the vast snow covered expanse it flew over.

This whitewash was later only applied to specific areas in the form of stripes or other larger patterns (see the Fw 190 F-8 kit) and was removed when spring came.

This too could be subject to modifications, with colours not really fit for the Russian Front such as RLM 79/80 from the Mediterranean area.

Southern Europe

The Fw 190 was not operational for long in North Africa. The planes that did operate in this theatre were firstly camouflaged

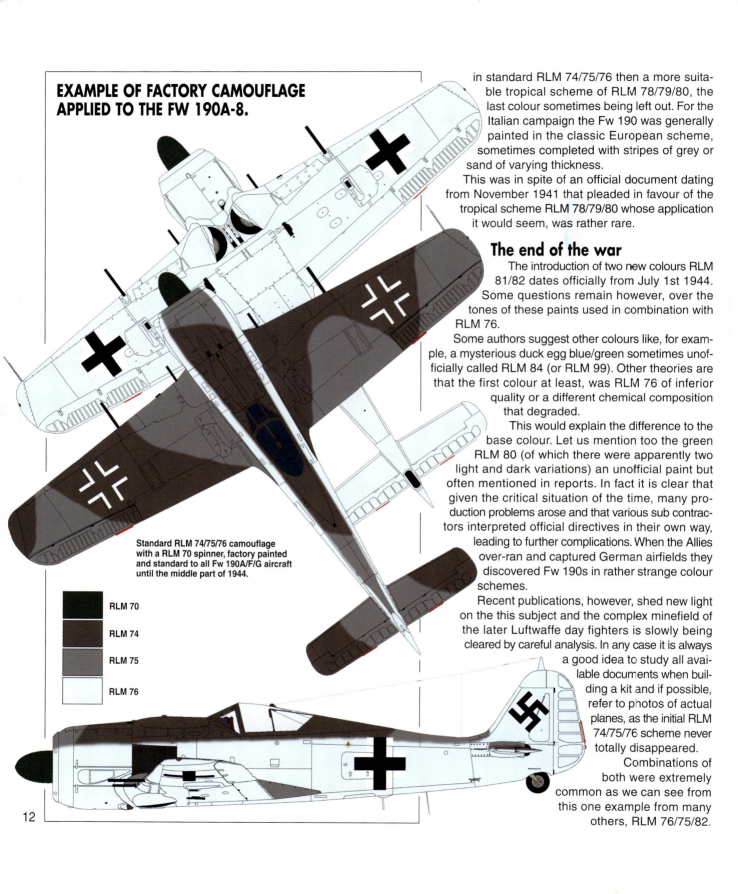

EXAMPLE OF FACTORY CAMOUFLAGE APPLIED TO THE FW 190A-8.

Standard RLM 74/75/76 camouflage with a RLM 70 spinner, factory painted and standard to all Fw 190A/F/G aircraft until the middle part of 1944.

- RLM 70
- RLM 74
- RLM 75
- RLM 76

in standard RLM 74/75/76 then a more suitable tropical scheme of RLM 78/79/80, the last colour sometimes being left out. For the Italian campaign the Fw 190 was generally painted in the classic European scheme, sometimes completed with stripes of grey or sand of varying thickness.

This was in spite of an official document dating from November 1941 that pleaded in favour of the tropical scheme RLM 78/79/80 whose application it would seem, was rather rare.

The end of the war

The introduction of two new colours RLM 81/82 dates officially from July 1st 1944. Some questions remain however, over the tones of these paints used in combination with RLM 76.

Some authors suggest other colours like, for example, a mysterious duck egg blue/green sometimes unofficially called RLM 84 (or RLM 99). Other theories are that the first colour at least, was RLM 76 of inferior quality or a different chemical composition that degraded.

This would explain the difference to the base colour. Let us mention too the green RLM 80 (of which there were apparently two light and dark variations) an unofficial paint but often mentioned in reports. In fact it is clear that given the critical situation of the time, many production problems arose and that various sub contractors interpreted official directives in their own way, leading to further complications. When the Allies over-ran and captured German airfields they discovered Fw 190s in rather strange colour schemes.

Recent publications, however, shed new light on the this subject and the complex minefield of the later Luftwaffe day fighters is slowly being cleared by careful analysis. In any case it is always a good idea to study all available documents when building a kit and if possible, refer to photos of actual planes, as the initial RLM 74/75/76 scheme never totally disappeared. Combinations of both were extremely common as we can see from this one example from many others, RLM 76/75/82.

FOCKE WULF 190 V-1
THE SAGA BEGINS

Being a prototype, the kit has a smooth and polished finish. This aircraft accomplished a whole series of tests in this configuration until January 1940 its code was changed to FO-LY in the summer of 1939. In January the cowling was changed and it received a new cooling system.

1/48 SCRATCHBUILT

WHEN THE TIME CAME TO BUILD THIS MODEL (not commercially available) I thought a lot on how to go about it given that there was not much information on this plane available. I started by thinking of transforming a shop bought kit but soon gave up because of the major differences between the prototype and production models.

In the end I decided to go for the scratchbuilt option but unsure if it should be a full model or obtained via the vacuform process. Choosing the latter meant making a matrix and a female part to join onto the original. I finally chose a standard master, using the plans from the Japanese Burindo magazine, number 78, that covers the Fw 190. So, it is a long, painstaking job that I invite you to discover in the following pages with detailed descriptions and pictures.

1. FUSELAGES AND WINGS

1 and 1bis. We begin by sculpting the fuselage and wings from a block of plaster using plans to guide us so that dimensions are correct. This is carried out by using files and various blades as well as different grades of sandpaper. Do not forget to consult any photos.

2. The next step is to make sure that all the parts fit together before moving on to the next stage. Any faults in the wings are filled with putty which also helps to reinforce the part. The fuselage is dealt with in the same way. After this the parts are varnished before making a resin copy.

3. As we can see here the original and the copy are identical. For this we used a silicon mould from which we cut out the lower section so that the part could be more easily extracted.

4, 5 and 6. The rounded shape of the nose is made using purpose built tools making sure that the inner diameter is correct. When this is done we sand the fuselage and wings with sandpaper and water.

7. Before making the copy and in order to make the next stages easier it is preferable to use a blade to hollow out the area corresponding to the cockpit.

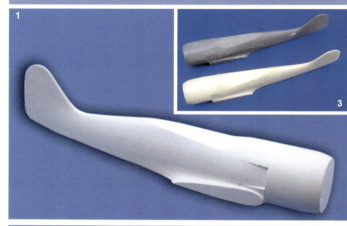

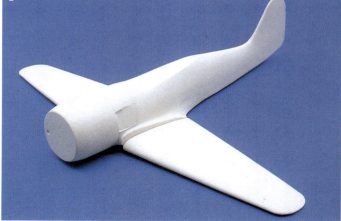

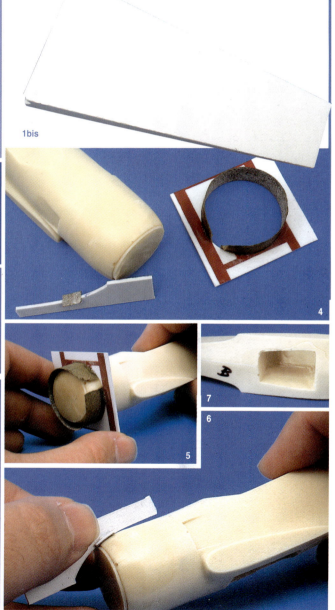

2. THE PROPELLER SPINNER AND THE PROPELLER

1 and 1bis. The spinner is made from a resin block using a mini drill and purpose built sculpting tool.

2. The prop spinner, after sculpting and the extremely simple metallic engraving tool.

3 and 4. The central line is made by using a worn blade and a mini drill. The central opening is hollowed out with a round drill bit.

5 and 6. The inside of the spinner is hollowed out in the same way as the outside. We finish off the part with a resin copy of the Tamiya engine and VDM propeller blades from the Otaki Fw 190A-8, longer and narrower so that they conform to the original.

7. The spinner is finished by adding the slightly inclined oblong openings as well as the interior spinner which is also scratch built.

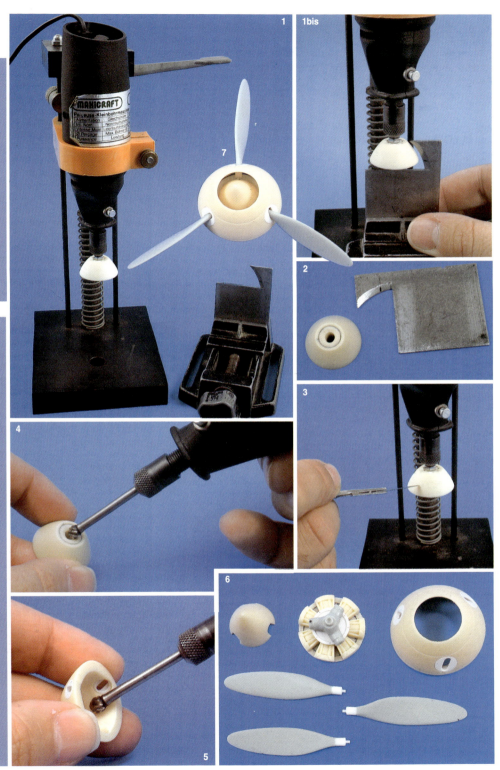

THE PARTS USED

Materials, parts and accessories necessary for the scratch building of the Fw 190 V-1

Fuselage. Plaster, silicon and resin
Engine. A resin copy of the Tamiya Fw 190A-3 engine
Propeller blades. Fw 190 A-8 by Otaki
Cockpit. Resin copies of the cockpit tub, seat, exhausts and instrument panel from the Tamiya Fw 190 A-3
Wheels. CMK accessories for the Fw 190 A-3
Other. Evergreen strips

THE COLOURS USED

INTERIOR
Instrument panels. RLM 66
Tub. RLM 02

ENGINE
Black Tamiya XF-1
Undercarriage, tyres, wells. RLM 02
Undercarriage legs, hydraulic system. Matt aluminium Tamiya XF-16 (enamel range)

WHEELS
Rim interior. RLM 02 - **Exterior.** Dark grey Tamiya XF -24. **Tyres.** Black Tamiya XF-1.

CAMOUFLAGE
Lower surfaces. RLM 65
Upper surfaces. RLM 70/71
Propeller spinner. RLM 70,
Internal spinner. RLM 02,
Propeller blades. RLM 70
Registration number. Black Tamiya XF-1
Swastika. Red RLM 23, white XF-2, black XF-1
Varnish. Brilliant Tamiya XF-2
Mattlack. Marabu (for the matt base)

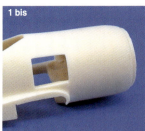

1 1 bis

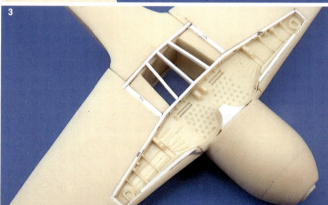

2

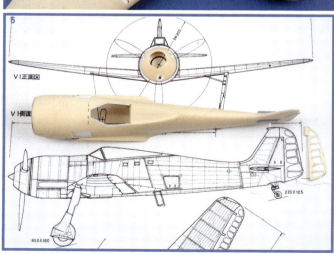

3

3. THE FUSELAGE AND WINGS (continued)

1 and 1bis. **We continue by cutting out the bottom of the cockpit, the nose and the exhaust outlets. This is done using a mini drill, a round drill bit and a curved blade.**

2, 3 and 4. **We also make an emplacement in the wings for the undercarriage wells, these are a resin copy of the Tamiya Fw 190 A-3. We then check, temporarily, the cockpit tub emplacement and that of the instrument panel with the fuselage.**

6 and 7. **Before moving on it is essential to compare the parts to the Burindo number 78 plans (Fw 190).**

8 and 9. **Before beginning on the detailing we check that the wings and propeller spinner are correctly positioned on the fuselage. This also applies to the plastic sheet covering part of the belly. The undercarriage wells are similarly checked.**

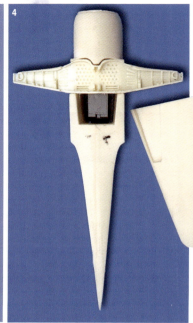

4

5

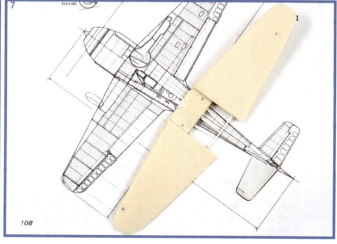

6 7

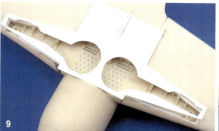
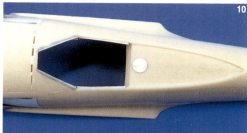
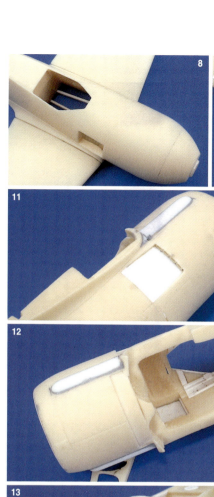
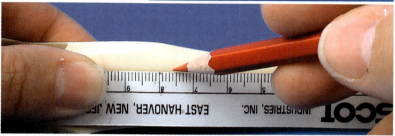

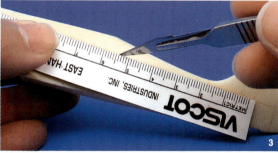

10, 11 and 12. We hollow out the area corresponding to the rear of the seat before placing the exhaust pipes and canopy rails.

13. All that remains is to check alignment of the spinner and engine.

4. THE HATCH LINES

1 and 2. We begin by working on the fuselage, then the wings by pencilling in the hatch outlines.

3. We go over the lines with the edge of a blunt scalpel, guided by a small metal ruler. As it is a resin kit, it is essential to remember that there could be small bubbles liable to be exposed by the scalpel.

4. A template is made from plastic card to ensure regular wing fillets.

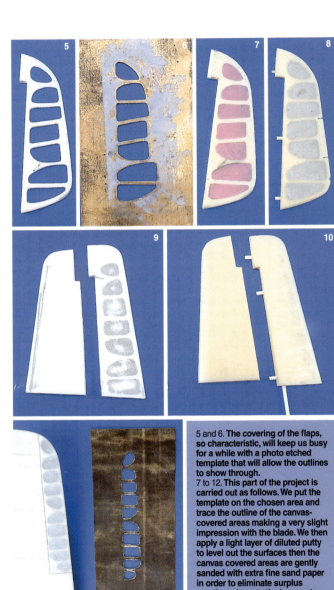

5. THE COCKPIT

5 and 6. The covering of the flaps, so characteristic, will keep us busy for a while with a photo etched template that will allow the outlines to show through.

7 to 12. This part of the project is carried out as follows. We put the template on the chosen area and trace the outline of the canvas-covered areas making a very slight impression with the blade. We then apply a light layer of diluted putty to level out the surfaces then the canvas covered areas are gently sanded with extra fine sand paper in order to eliminate surplus mastic and create the impression of tension to the canvas.

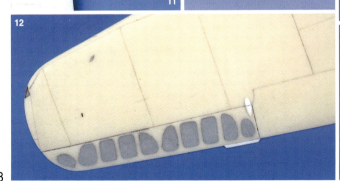

1 and 1bis. We begin with the Tamiya Fw 190 A-3 instrument panel by cutting the upper panel to reduce the space separating it from the lower panel.

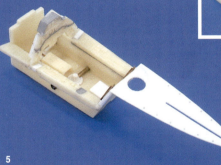

2. Evergreen is used next to make the rudder pedals, joystick and canopy opening lever.

3. The only seat detailing is in the seatbelt. This is made from copper sheet.

4, 5 and 6. We finish off with the cockpit tub detailing. A seat-positioning handle is added along with throttle lever, cables and the rear cabin extension (this is made from Evergreen plastic). These various cockpit parts are ready to be painted.

6. THE CANOPY

1. The canopy is made using the thermo-moulding technique, this is simple to do and cheap. This does, however, require a certain level of skill and a careful approach.

2. We begin with the mould in two parts (male and female for the smallest). These are made to the exact dimensions of the part. It is essential to use heat resistant, hard materials.

3. The sheet of clear plastic is then fixed to a wooden frame using aluminium wing nuts. We then heat it with a thermal scraper, any other heat source can do this as long as it is not gas (this will stain the plastic), for example a hairdryer can be used.

4. The male part of the mould is then slid under the plastic as it starts to soften. Continue applying the heat but not in the same place as it will burn the transparent sheet.

5. The female part, similarly treated, is placed on the male part whilst pressing lightly so that the two mould parts are joined before leaving it to cool.

6. The canopy is then extracted from the mould so that it can be carefully cut and detailed.

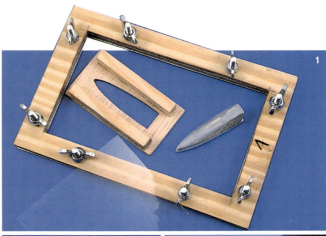

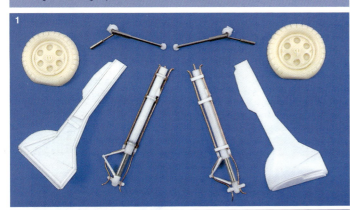

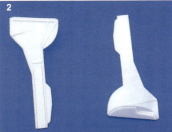

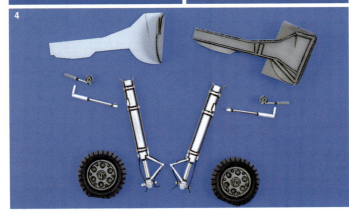

7. THE UNDERCARRIAGE

1. The undercarriage legs are made from Evergreen tubes with the insides reinforced by Minemeca metal rods. The undercarriage and the CMK wheels await painting. Note the copper sheet and wire detailing.

2. The undercarriage doors are next; they are made according to the plans from 0.3 mm Evergreen plastic.

4. The tail wheel gear is partially made with plastic tubing. The wheel itself is made from a thick piece of plastic with the help of a mini drill.

5. A view of the undercarriage after painting. The legs and hydraulics are in aluminium with the doors in RLM 02 for the interior and RLM 65 for the exterior.

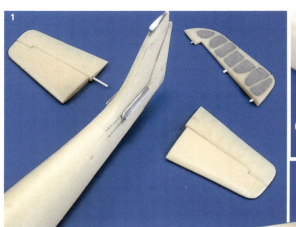

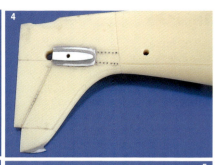

8. ASSEMBLING
AND PAINTING THE OTHER PARTS

1, 2 and 3. The cockpit tub is installed in the fuselage. This is followed by cementing the wings, tail plane, rudder and the aerial mount.

4. The adjustable tail plane fillets remain to be made out of plastic sheet.

5 and 6. The initial phase of the cockpit assembly begins by painting it in RLM 02.

7 and 8. The cockpit tub, instrument panel and rudder pedals are done in the same way then assembled.

9, 10 and 11. The engine and interior of the spinner are painted before being fitted. The top of the fuselage after assembly and a close up of the spinner ready for assembly.

12. We finish off by painting the wells and installing the plastic sheet central panel upon which we must not forget to etch the structural lines.

The plane is ready for painting.

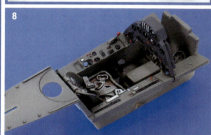

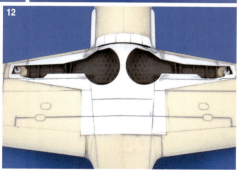

THE PLANE'S CAMOUFLAGE

The prototype, first flown by Hans Sanders on June 1st 1939, was initially entirely painted in RLM 02 then camouflaged in RLM 65/70/71 in accordance with German fighter planes of the time. Working out the exact camouflage scheme on the wings, lower surfaces and especially the left side, was not an easy task. The photo of the actual plane only showed the right side and it took hours and hours of watching a video of the Fw 190 taken from the series « *planes of the Third Reich* » frame by frame to get an exact idea.

---

1. Given the origin of this study, we do not know if this video is available in France and if it is the title could be different.

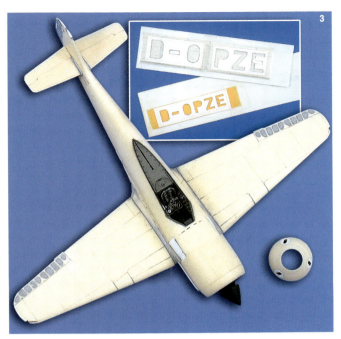

1. From the two possible paint schemes, the standard 1939 camouflage was chosen. We start with the tail fin insignia, a white circle is painted then masked off so that the red can be applied. The swastika is then painted on using a homemade stencil.

2. The wells, tail, nose and cockpit are masked before beginning the camouflage, which begins on the belly with RLM 65. All theses lower surfaces are then masked (up to the leading edge of the wing). The paint is then applied using stencils for the correct pattern and in the following order, RLM 71 then RLM 70.

3. The registration number is painted on using stencils cut out from slightly adhesive Tamiya tape.

4. A coat of gloss Tamiya XF-22 varnish is applied. The code letters are done next in dark grey for the lower wing surfaces and black for the upper.

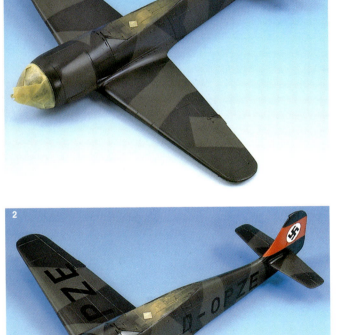

Close up of the tail and tail fin, allowing us to see the tail wings, tail wheel and stencilled flag.

The finished model seen from above, showing all of its shape, thin tail and straight wings and their rounded wingtips which disappeared from the production models.
The wind speed boom, made with a Minemeca metal rod and the registration number. Close up of the cockpit showing part of the seat and the one-piece canopy. In spite of the dark camouflage, the rectangular tail wing and its particular canvas covering stands out quite well.

The belly differs from the other Fw 190s with its ring of cooling flaps in front of the wells. Other differences are in the undercarriage doors and the position of panels and maintenance hatches, not forgetting the larger nose and more streamlined stabilizers.

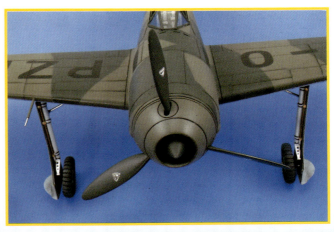

Above and below.
Close up of the cockpit and the carefully finished canopy joins and in particular the windscreen. The canopy is cemented to the fuselage using white glue. Note the dark colour of the exhaust panel. Partial view of the code on the prototype's right side.

Above.
The logo on the propeller blades is a transfer made using specialist CorelDraw software, which is then transferred onto an acetate sheet not destined for printers, to stop the ink being absorbed. We then gently dry the logo using a hair dryer and leave it for several hours. After that, it can be placed where needed by using a pencil to rub the reverse side. We advise you not to move the logo over the propeller blade, as the ink is unstable and could result in staining. Seen from this angle, the nose, spinner and undercarriage show its distinctive position at a reflex angle to the wing seen from the front.

Below.
This aircraft continued with tests throughout 1940 along with four other prototypes, the V-2, V-3 (an unfinished machine destined to supply spare parts) V-4 and the V-5.

THE ALLIED BOMBING CAMPAIGN that hit industrial centres in daylight raids and civilian targets at night, progressively reached alarming proportions with an ever-increasing number of American planes over the Fatherland from 1943.

One of the first lines of defence was composed of fighters based on the French and Dutch coast. Apart from their usual missions they were also allocated the task of intercepting the B-24 Liberators and B-17 Fortresses as they flew to and from their targets. This task fell to JG 1, JG 2 and JG 26 who attacked the allied heavy bombers with relative success. The arrival of the Mustang and its role of escorting the bombers to and from the target led to a decrease in losses to the German fighters.

The German fighters' task of infiltrating the tight bomber formations was far from easy as they became vulnerable to the defensive fire of the American bombers.

This added to the stress that these men were subjected to and there were often periods of discouragement, so frequent when chasing results. The shooting down of a four-engined bomber counted for more than a fighter when it came to handing out medals.

The pilots of JG 1 who operated in the same combat zone as the P-47 and P-38 flew spectacularly decorated planes with chequered cowlings or large black and white bands. It is one of these Focke Wulfs that will be the theme for our next project, starting off with the excellent Tamiya kit.

Making this Fw 190 A-4 required few accessories but some, like the Aeromaster decals turned out to be counter productive.

Transformations and improvements

Making a Fw 190 A-4 from the Tamiya A-3 version kit is fairly straightforward as only a small number of details need to be added.

We decided, however, to improve it further by leaving the wing and tail flaps down and the undercarriage doors open as well as the tail wheel mechanism hatch. Decorating this spectacular JG 1 plane, with its black and white cowling, seemed fairly easy at the beginning, especially as a chequered decal sheet by Aeromaster was available.

As it turned out, opting for this decoration made for extra work, as we will see in the relevant chapter. A CMK set will allow us to replace the mobile surfaces whilst an Eduard photo etched set will, as usual, enable us to make the flaps and improve the cockpit. To begin we need a good blade for the mobile surfaces including, of course the flaps. We do the same with the central undercarriage doors that are fixed to the wings, cutting along the lines in the plastic.

We then open the triangular tail wheel hatch, once the rudder has been removed, being extremely careful with its external edge that will need to be thinned down from the interior so that it conforms to the scale. This done, the CMK resin parts of the tail wheel retraction mechanism are fitted. We next make the system spring from 0.1 mm copper wire.

This is followed by compensating the thinning down of the inner side of the left tail fin half by reinforcing the resin part with tiny plastic strips around the tin housing on the long shock absorber. Straight after we paint the compartment interior and the hatch in RLM 02 for the metal parts and in buff for those covered with canvas. We apply a light wash and a few highlights with an acrylic Prince August paint close to RLM 02 but mixed with white.

Closing the fuselage we note that the CMK rudder is narrower than the kit's fin so we go back to the finished original part whose leading edge will need to be re-profiled using plastic sheet in order to regain its natural curve.

Let us see what the undercarriage has in store for us. The

(continued on page 32)

FOCKE WULF 190 A-4
« *CHECKMATE* »

1/48 scratchbuilt

The Focke Wulf Fw 190 A-4.

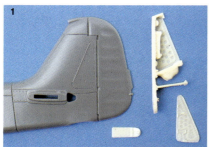

1. RUDDER, TAIL, WHEELS

1. The parts used for the tail wheel retraction system come from a resin CMK kit (ref. 4025)

2 and 3. The CMK kit includes two types of rudder, one of which, at first glance seemed to correspond to the first versions of the plane.

4, 5. The elevator removed. We thin down the internal part of the tail fin, which unfortunately makes the CMK, part too narrow to be used.

6. The cockpit sides are thinned down. The work in progress allows us to see the difference between the two fuselage halves.

7. The CMK tail wheel retraction kit is completed with wound round copper wire.

8. Before closing the fuselage we dry fit the resin parts. The fit is not perfect due to the thinning down so we add small plastic bands around the shock absorber insert.

9. Everything is painted in RLM 02 except for the tin part in buff; we then add the washes to highlight the contours.

10. he Tamiya tail wheel is too small and is replaced by the more detailed CMK wheel.

11. The differences between the Tamiya, CMK and Trimaster tail wheels are obvious. Seen from left to right.

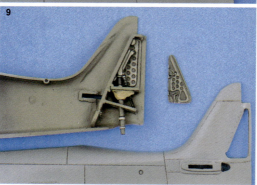
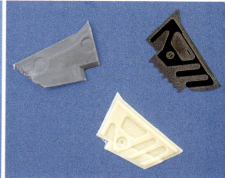

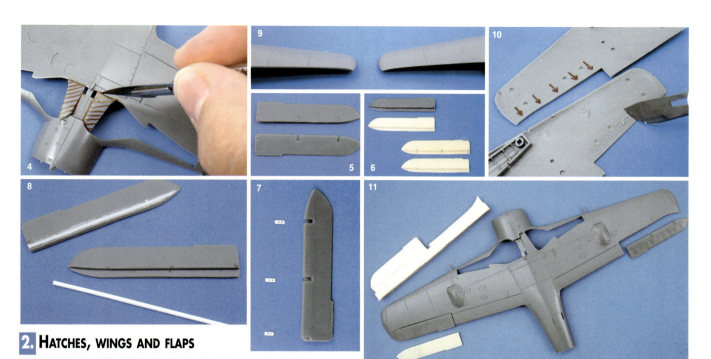

2. HATCHES, WINGS AND FLAPS

1. The lightly detailed Tamiya hatch can be replaced by the metal photo etched Eduard part (lacking in volume) or by that of Cutting Edge included in a rather expensive kit. You can also, as here, opt for the scratch built solution.

2 and 3. The building of the hatches from plastic sheet and a two-component mastic.
The photo etched Eduard part serves to add overall detail but not volume. The work finished, all that's left to do is to make a resin copy.

4. The internal hatches are marked with a black felt tip before being cut out and replaced with new parts, which will remain open.

5,6 and 7. The flaps and ailerons are cut out so they can be placed in a down position. Three notches are cut into the ailerons in order to fit plastic sheet hinges.

8. The front part of the ailerons is rebuilt with Evergreen plastic.

9 and 10. The wing is thinned down for the ailerons, the improvement is obvious when you compare the two photos.

11. The Cutting Edge resin parts are perfect and slightly better than those of Tamiya.

12. The flaps, marked in felt tip, are cut out with the help of a blade.

13. The photo etched Eduard flaps are lacking in volume compared to the originals and some work on them is needed.

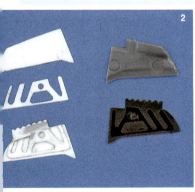

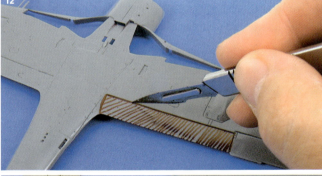
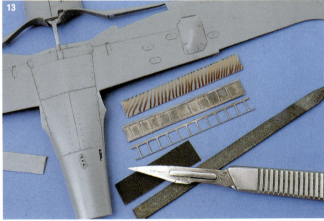

The right wing of the FW 190 A4.

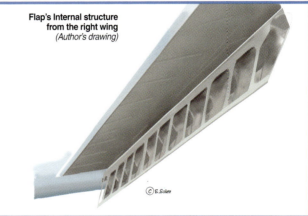

Flap's Internal structure from the right wing (Author's drawing)

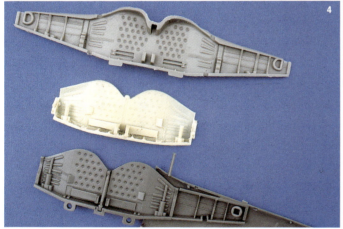

3. UNDERCARRIAGE, HATCHES, WINGS AND FLAPS

1. We therefore make thicker flaps by adding plastic sheet and soft aluminium to the Eduard parts.
2. One of the finished flaps and the different parts that went into its construction.
3. The CMK insert of the undercarriage wells is more detailed than the original. The cannons were not interdependent of the bottom of the bays.
4. Eduard photo etched transverse reinforcing and part of the engine mount tubes are added. The undercarriage well detailing differs from one manufacturer to another. Trimaster for example has a one-piece part whilst Tamiya and CMK's have a separate central part.
5. It is painted in RLM 02 and has highlights added as well as washes for tone and relief
6 to 7. Undercarriage details on a Fw 190 A. The interior, wrongly painted in blue, lacking certain elements, the cannon does not have its leather protection.
8. We next add the hydraulic system using rods and copper wire.
55 to 64. Various photos of the undercarriage showing the torque link, inside of the hatch and the main drag strut with its electric connections and bolts on the external side.
9. Several types of wheel were mounted on the Fw 190. Above the first cut out models and below, the last with their bolted wheel hub covers
10. The Tamiya wheels (in grey) are much too small. Those made by True Details (of a yellowish colour) and CMK are of a much better size.
11. The undercarriage legs are improved by removing the thicker parts like the torque links and by adding Eduard photo etched parts.
12 and 13. The last thing to do is lower the elevators once they have been cut out.

(Note. This is not an absolute necessity given the actual thickness of the Fw 190's torque links.)

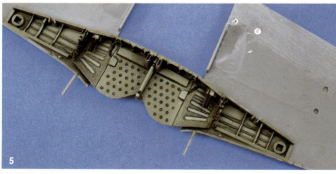

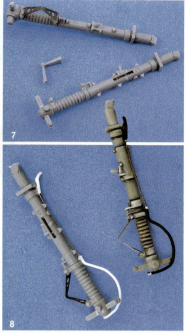

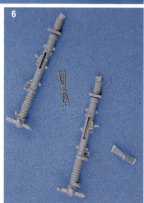

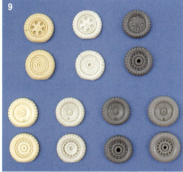

COLOURS USED

INTERIOR
Cockpit RLM 66 (Gunze Sangyo/Prince August)

UNDERCARRIAGE AND TYRES
Undercarriage, wells and interior hatches. RLM (Prince August ref. 886)
Tyres. grey black (Prince August ref 862)

CAMOUFLAGE
lower surfaces. light grey blue RLM 76
Upper surfaces. grey RLM 74/75
Tactical markings. yellow RLM 04
Propeller spinner. white
propeller blades. Green black RLM 70

VARNISH
Putting on the decals. Tamiya gloss acrylic.
Finishing touches. Marabu synthetic matt

Note that the camouflage and some of the more intricate work was done using airbrush applied acrylics by Gunze Sangyo/Tamiya. All the mixing was based on the Federal Standard colour chart.

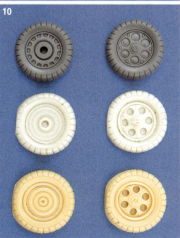

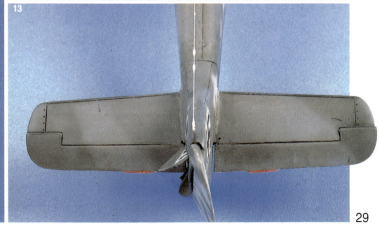

29

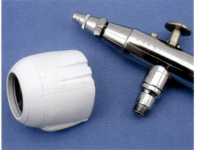
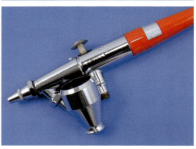

Close up of the different parts of the finished model cockpit, canopy, cowling, guns, etc.

The chequered cowling.

The plane the model is based on after an undercarriage malfunction. We can clearly see the previously mentioned particularities such as the dark colour of the cross and the density of the camouflage.
(RR)

PAINTING THE CHEQUERED COWLING

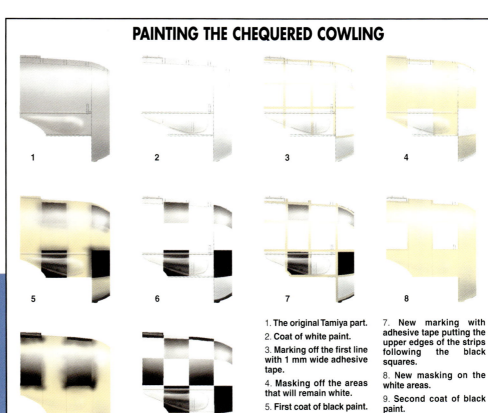

1. The original Tamiya part.
2. Coat of white paint.
3. Marking off the first line with 1 mm wide adhesive tape.
4. Masking off the areas that will remain white.
5. First coat of black paint.
6. The initial masking is taken off.
7. New marking with adhesive tape putting the upper edges of the strips following the black squares.
8. New masking on the white areas.
9. Second coat of black paint.
10. The end of the process

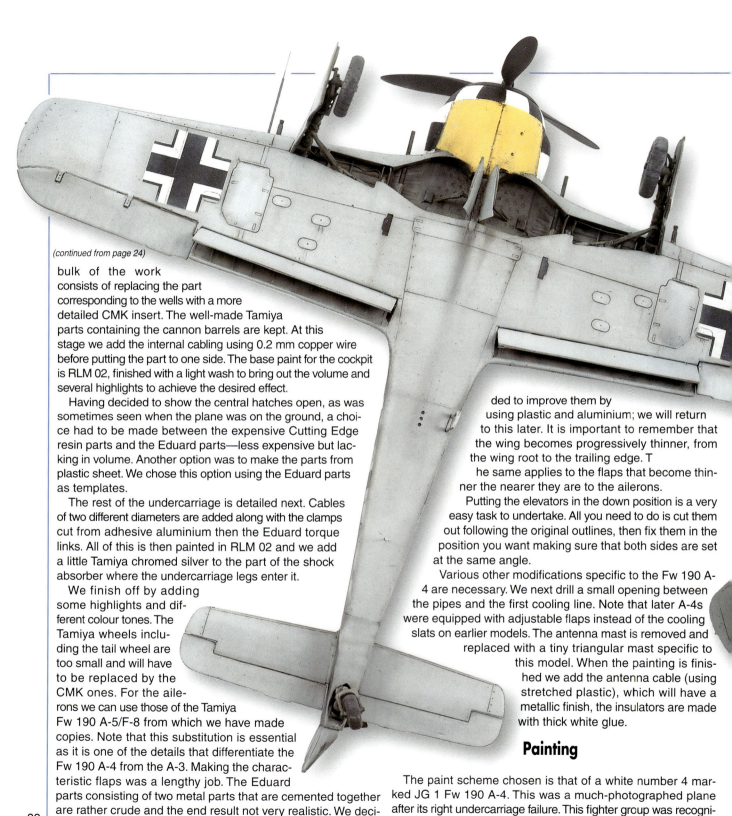

(continued from page 24)

bulk of the work consists of replacing the part corresponding to the wells with a more detailed CMK insert. The well-made Tamiya parts containing the cannon barrels are kept. At this stage we add the internal cabling using 0.2 mm copper wire before putting the part to one side. The base paint for the cockpit is RLM 02, finished with a light wash to bring out the volume and several highlights to achieve the desired effect.

Having decided to show the central hatches open, as was sometimes seen when the plane was on the ground, a choice had to be made between the expensive Cutting Edge resin parts and the Eduard parts—less expensive but lacking in volume. Another option was to make the parts from plastic sheet. We chose this option using the Eduard parts as templates.

The rest of the undercarriage is detailed next. Cables of two different diameters are added along with the clamps cut from adhesive aluminium then the Eduard torque links. All of this is then painted in RLM 02 and we add a little Tamiya chromed silver to the part of the shock absorber where the undercarriage legs enter it.

We finish off by adding some highlights and different colour tones. The Tamiya wheels including the tail wheel are too small and will have to be replaced by the CMK ones. For the ailerons we can use those of the Tamiya Fw 190 A-5/F-8 from which we have made copies. Note that this substitution is essential as it is one of the details that differentiate the Fw 190 A-4 from the A-3. Making the characteristic flaps was a lengthy job. The Eduard parts consisting of two metal parts that are cemented together are rather crude and the end result not very realistic. We decided to improve them by using plastic and aluminium; we will return to this later. It is important to remember that the wing becomes progressively thinner, from the wing root to the trailing edge. T he same applies to the flaps that become thinner the nearer they are to the ailerons.

Putting the elevators in the down position is a very easy task to undertake. All you need to do is cut them out following the original outlines, then fix them in the position you want making sure that both sides are set at the same angle.

Various other modifications specific to the Fw 190 A-4 are necessary. We next drill a small opening between the pipes and the first cooling line. Note that later A-4s were equipped with adjustable flaps instead of the cooling slats on earlier models. The antenna mast is removed and replaced with a tiny triangular mast specific to this model. When the painting is finished we add the antenna cable (using stretched plastic), which will have a metallic finish, the insulators are made with thick white glue.

Painting

The paint scheme chosen is that of a white number 4 marked JG 1 Fw 190 A-4. This was a much-photographed plane after its right undercarriage failure. This fighter group was recogni-

The finished model seen from above and below.

zable for a while by its superb chequered cowlings, of which the colours, black and white, black and red or black and yellow, varied according to squadron. Knowing that Aeromaster made a decal kit we bought it but what a let down! The layout of the black and white squares differs from the photos. As well as this the IV./JG 1 badge, depicting a little devil in the clouds that is badly drawn and two are provided when only one was painted on the port side of the plane. The crosses, except those of the lower wing surfaces, were also disappointing. All that was usable were the white numbers. We therefore decided, considering everything that needed redoing, to leave these decals and get on with painting this superb and elegant decoration ourselves.

We made the number four from an adhesive film stencil. We did not, however, add the black outline despite what we can see in certain profiles. We decided to follow what we could see from the photos. For the upper wing crosses we used both stencils and Eduard decals. The devil badge is hand painted whilst the other decals are by Tamiya.

A close-up examination of the camouflage, using the photos at our disposal, allows us to add certain very interesting particularities. It seems at first to be a plane with the camouflage scheme RLM 74/75/76 typical of the time, but the fuselage scheme is fairly unusual for a day fighter. Indeed we can see that whereas the front has a standard configuration with a demarcation line between the grey and the RLM 76 situated at about cockpit height which is slightly mottled on the sides, the camouflage gets darker towards the tail and the latter is painted mainly in RLM 75 and 74 with just a small stroke of RLM 76 for the swastika. This particularity has led several authors to believe that this plane, at some point in its operational life was used for night missions. This is somewhat backed up by the fuselage cross where a grey (probably RLM 77) replaces the white. The sides behind the exhausts, were stylistically painted in black as an elegant way of hiding the exhaust trails. It is possible, judging from its rather imprecise outline against the camouflage, that these were frequently repainted. The cowling chequers are airbrushed with white *Prince August* and black Gunze paints. The yellow band underneath is added last.

Two realistic views of the finished Focke Wulf 190 A4.

The finished model seen from the left side.

FOCKE WULF 190 A-5

« THE SUPERIORITY OF THE BMW ENGINE IN EXTREME COLD »

Photo of the finished model showing the BMW 801 D-2 engine with its twin row 14 cylinders cooled by a ten bladed fan. It developed 1700 hp at take off and up to 1440 hp at 5800 m.

THE FIRST SIGNIFICANT batch of Fw 190A-5s appeared in February - March 1943. It was the result of an improved production line (improvements implemented from the end of 1942) of which the earlier versions were in dire need. The most important change was the BMW 801 D-2 engine that meant lengthening the fuselage by 15.25 cm from the wing junction to the engine cowl. The A-5 was also designed with a large number of «Rüstsätze» in mind leading to the addition of a letter R to the original version code. From now on the Fw 190 became the object of a host of developments and factory prototypes destined for service on various fronts.

The extreme attention to detail, so typically German, of the Luftwaffe's system of numbering, is extremely useful to modellers. By studying the abundance of available documents we are able to find some lesser-known versions, which result in some highly original models, especially as specialist shops stock plenty of accessories and decal sheets. We decided to choose one of the planes used by *Leutnant* Emil « Bully» Lang, one of the Luftwaffe's greatest, and sadly lesser-known pilots.

Emil «Bully» Lang joined the famous Fw 190 equipped JG 54 in October 1942 while the unit was based in Russia. Lang was a robust, athletic man who acquired the nickname Bully due his bulldog like appearance… he very quickly notched up many kills, reaching 73 by October, including 12 the same day!

By December 1943 his total number of kills was 145 with 18 planes shot down the same day, thus beating the 17 kills of H.J. Marseille in North Africa. He was later transferred to the western theatre with the rank of *Gruppenkommandeur* to 11/JG 26. Still flying an Fw 190 he continued his tally of kills. He shot down three Spitfires in five minutes on July 9th 1944, two P-47s in one minute on August 15th, two P-38s on the 25th of the same month and finally three Spitfires on the 26th.

He was, however, shot down by a Thunderbolt over Belgium on September 3rd after a technical problem stopped him from retracting his undercarriage. His total number of kills was 173.

The kit we used was the near perfect Tamiya. Being the F-8 version it differs slightly from the F-5 we wanted to build so we needed to make certain modifications. We should tell you that a fairly good quality Dragon Fw 190A-5 could do the job although it is not always an easy kit to find.

A Cutting Edge conversion will be extremely useful to make this Fw 190A-5. Its numerous advantages include not only its excellent compatibility with the Tamiya kit but also the fact that the A-6/A-7 and A-8 versions can also be made using the Tamiya kit as a base. The armament was also completely overhauled using Verlinden/Aires kits that supply the wing cannon bays.

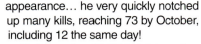

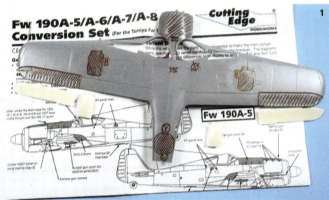

1. The Tamiya Fw 190F-8 surrounded by various resin and photo etched parts.
2. The lower surfaces are marked with a felt pen where the Cutting Edge parts will eventually go or modifications made.
3. The cannon access hatch is drilled out all around its edge to make it easier to cut out.
4. The aileron will be replaced by a resin part after it has been cut out with the help of a triangular blade.

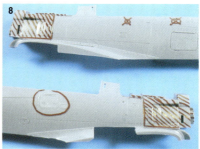
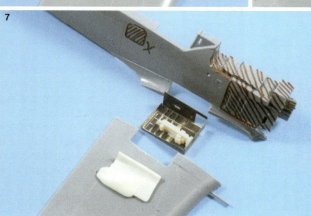
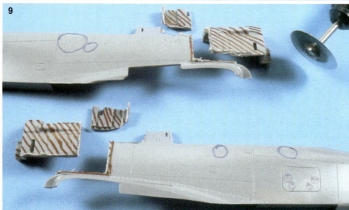
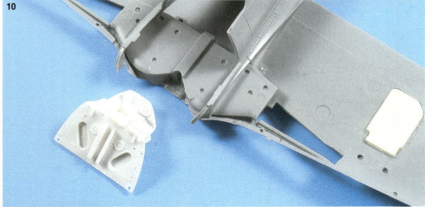

1. Various transformations

4. The panels specific to the A-5 are cemented with cyanoacrylate, which will also double up as putty when sanding.
5, 6. The modified areas corresponding to the cannon bays are coloured with a red felt tip then cut out using a circular saw mounted on a mini drill.
7. We make a first trial run installing the Aires/Verlinden parts then we file and sandpaper until the assembly is perfect. Note the red marks on the fuselage-half indicating the areas to remove.
8. We then cut out the marked areas on the fuselage halves. Other details that are not present on the Fw 190A-5 such as the starboard radio transmitter maintenance hatch, the port, central and auxiliary refuelling ports are marked off in red, refilled with cyanoacrylate then sanded.
9. The fuselage halves during the cutting out phase and the forward cowling partially raised. The areas circled in blue show the modified panels and hatches.
10. The fuselage is now assembled and the edges of the cut out areas are smoothed off as well as possible from the inside. We can do a dry fit test to make sure that the parts fit together well and if need be make any changes, especially at the firewall that will have the jig and engine fixed to it.
11. Trial and installation of the MG 151 20 mm cannon bays
12. The undercarriage wells and the MG 151 cannon bays are brush painted using Prince August acrylic paints, then given an airbrush patina with a very diluted black mixture.

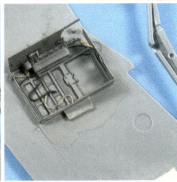

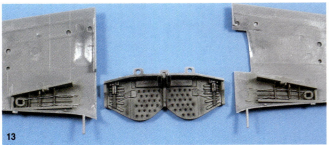

13.

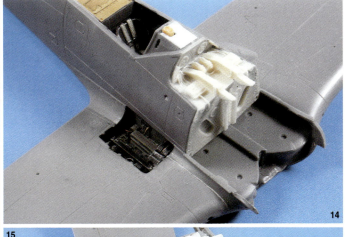

14.

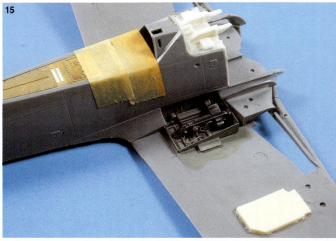

15.

I

II

III, IV

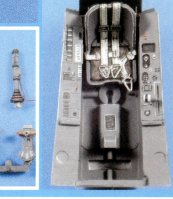

V

The cockpit interior is composed of a resin and photoetched CMK kit, an Eduard sheet as well as certain parts kept from the kit such as the cockpit tub, seat and joystick.
We have also used Cutting Edge parts for the undercarriage doors and the closing mechanism. This is completed with True Details wheels destined for the A-1 to A-5. The other parts come from the kit; they have just been improved by the brake circuit on the undercarriage legs and other various minor details.
I. The Tamiya, CMK and Eduard parts for the cockpit are ready for painting. The white plastic part on the left is for supporting the photo etched panel just behind the cockpit.
II. The cockpit is first placed temporarily on the left half of the fuselage so that any adjustments can be made. The second picture shows the plastic sheet support of the aft compartment used for the storage of personal items.
III, IV, V. The resin and photo etched instrument panel, brush painted with acrylic Prince August paints, as well as the tub, control column and pedals.

13. For all internal parts such as the bays, undercarriage, engine or cockpit we use brush applied Prince August acrylics. The metallic parts are brush painted in Tamiya XF-11 and XF-16 enamels and finished off by using the airbrush, where needed, to add shade and tone.
14, 15. We now assemble the modified fuselage and the wings (beginning with the lower surfaces) finishing off by installing the Revi gun sight followed by the windscreen.
16. The undercarriage is finished off with various resin parts. Everything is ready and awaits only painting. Note the characteristic shape of the secondary doors and their internal detailing.

16.

2. POWERPLANT

The CMK BMW 801 was specially designed for the Tamiya kit, which is a real bonus. Made from resin and photo etched parts it is a very complete kit but on the downside the instructions could do with being more precise. This lack of precision means that the modeller has to resort to studying documents, which are luckily readily avail-

(continued on page 39)

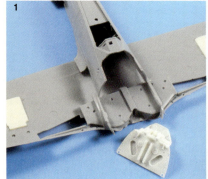

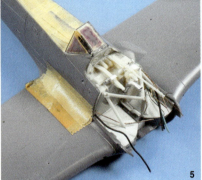

1, 2. The gap between the engine and the cockpit has the machine gun supports added. Luckily the assembly is perfect here.

3, 4. Plastic sheet is then added whilst the windscreen and cannon are masked off.

5. We then make a trial run of the engine mount, a very important stage as the assembly must be perfect. We next cement the various cables and tubes, machine guns and ammunition drums.

6. The engine bay is brush painted with Prince August acrylics followed by a light wash of very diluted Tamiya XF-1 that gives a grimy finish. This is applied with an airbrush.

7. We now proceed to the permanent fixing of the engine mount taking care to measure properly, then carefully add the machine gun drums. The rear part of the engine is then added having been painted beforehand.

8. The central engine part is fixed to the mount. Cables and tubes are added as well as other parts. We check that the machine guns are correctly aligned.

s. The addition of the central part of the engine calls for very precise positioning. Any mistake made at this point will lead to other parts being out of line and there is no going back. The following picture shows several PVC parts, cemented with cyanoacrylate around the reducer.

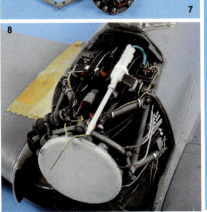

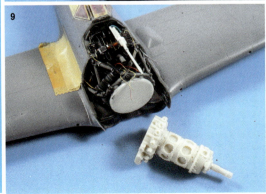

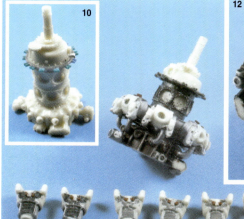

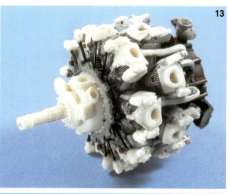

10,11. The central area and cylinders are now brush painted with Tamiya enamel and Prince August acrylics for the other parts. Note the gas tubes added to the rear of each cylinder.

12, 13. All the cylinders are now cemented along with the pipes and the front stems. Certain areas are not yet painted so that the engine can be handled more easily.

14. After having added the spark plug cables (two per cylinder) we finish off cementing the engine and add some more hose.

The building of this type of engine means alternating between construction and painting as many parts have to be gradually added.

15. We add the exhaust pipes as well as other various parts between the rows of cylinders. Some pipes need to be painted before cementing whilst others only after removing the masking tape.

16. The forward part of the engine to which we add various details such as tubes and cables which are painted as the work progresses.

able. Some of the more delicate parts are held in place by enormous stalks, which means a very long and patient job of separating them using various blades and saws so that they are not damaged.

The cowling mounted MG 17 7.92 mm machine guns come from the Aires kit, the resin used is extremely fragile and needs some home made strengthening. Engine improvements are finished off by adding copper cables, tiny PVC tubes, tin plate and strips of Evergreen plastic. This laborious and often stressful job will however result in a very spectacular BMW 801 with the panels open. This will really improve the overall presentation of the model.

3. CAMOUFLAGE

1, 2. The model is ready for the camouflage. All areas not concerned by this are masked off using Tamiya adhesive tape, tissue paper and Humbrol Maskol. From now on the painting will be mostly done with the airbrush.

3. We start off with the light grey blue

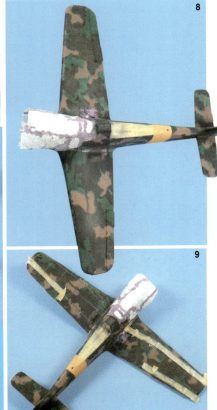

RLM 76, which is left to dry for an hour before painting the wing tips in RLM 04 yellow making sure that the rest of the wings are protected by tissue paper and adhesive tape.

4. Weathering and painting the panels with the airbrush using a very diluted grey Tamiya XF-24.
5. The inside is finished. The grey of the belly has been lightly painted very diluted blue RLM 76. All that remains is the oil wash and varnishing.
6, 7, 8. We next proceed to the plane's upper surfaces, adding the mottling in this order. Sand RLM 79, green RLM 25 and olive RLM 80. This is then followed by the markings on the fuselage and tail plane in yellow RLM 04.
9, 10. The wing and fuselage hatches are highlighted in light grey XF-55. This is done line by line with the help of adhesive tape, starting with the horizontal lines then the vertical ones.
11. Highlighting the canvassing on the ailerons and tail planes is done using adhesive strips laid along the ribs. It is a long job but the result is spectacular.
12. We now apply a coat of Micro Gloss (Microscale) varnish, followed by a wash of black oil paint to accentuate the panels (various recesses and slits) before removing any excess with the help of a brush lightly dipped in thinners.

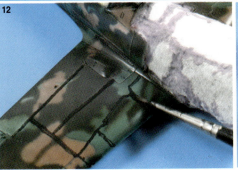

11, 12. The canopy, with its supports already painted, is fitted temporarily in order to check that it fits to the fuselage and the way it slides back. Note that the decals are already in place at this stage and protected by a coat of Marabu matt varnish. The masking on the undercarriage and engine can now be removed and certain small parts such as the machine guns fitted.

13. When we are sure that the canopy is perfectly positioned it is airbrushed and brush painted with a fine brush for the finer exterior details.

14, 15, 16. The cowling parts, ring and fan included, painted and awaiting fitting. The internal sides of the panels are in grey - green RLM 02, highlighted in black and, once dry, with pale paint.

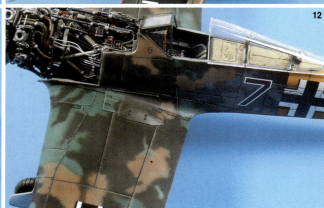

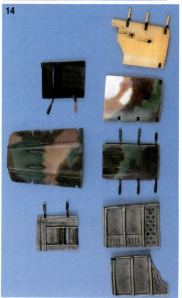

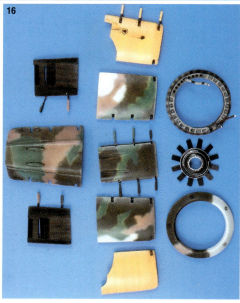

17. The propeller is painted in black green RLM 70 and is given a light airbrush applied weathering. The other parts are the forward armoured ring and part of the rear cowling, the cooling fan and its ring.
18. The finished model with its engine visible, by far the most difficult part of the assembly to master.
19, 20. Engine details and certain cowling panels placed on the wings and around the model.
21. The MG 151 cannon bay and its hatch show two interesting points. The electrical connections to the rear of the ammunition belt and the compartment surround painted in RLM 02 on the lower part.
22. The tail wings with the fake rib contours and the elevator lower surfaces. These are lightly highlighted with an airbrush applied dark tone.

THE PLANE'S CAMOUFLAGE

One of the first to receive the new Fw 190 version was without doubt Emil « Bully » Lang in the spring of 1943. The German army was preparing for « Operation Citadel » and the Luftwaffe airfield at Siverskaya, in the middle of the steppe, had important installations that were destined to play a vital role for the air support needed in the operation.

The plane's camouflage

It was here that the plane we are using for the model was photographed. The decals are made by Aeromaster under the reference 48432. This aircraft belonged to 5./JG 54 and judging from the period photos, those in charge decided to adopt a camouflage scheme that differed from the other Jagdgeschwader.

This unusual scheme, both beautiful and original was painted over the official grey livery that was deemed to be ill suited to the spring and summer colours of the steppe.

The paint used for these aircraft was always of a superior quality. We can see from many documents that there is little evidence of flaking. All factory markings have seemingly disappeared under this new camouflage that was certainly not yet officially acknowledged.

The RLM colours that make up the camouflage on our Fw 190A-5 should, in our opinion and according to the official guidelines of the time, be the following RLM 25 (light green), RLM 79 (light brown) and RLM 80 (olive green) for the upper surfaces and RLM 76 (light grey/blue) for the belly and sides. RLM 25 was a very beautiful green, usually used for the numbers but seen by some people as being used on a certain amount of Russian Front planes.

The RLM 79 and 80 colours were destined for the

NDE. Recent studies from K.A. Merrick and J. Wolf's book, *Luftwaffe camouflage and markings, 1933-45 vol.1* (Ian publishing 2005) hint that the following colours were used from the summer of 1943 to approximately the end of 1944 when the last planes disappeared from the front. RLM 61 *Dunkelbraun* (a dark brown from which RLM 81 derived) RLM 62 *Grün* (a medium green that led to RLM 82) and RLM 64 *Dunkelgrün* (an olivey green close to RLM 83). These colours were moreover combined with grey - green RLM 02, often applied as a flat tint in irregular and varied shapes.

3. THE CAMOUFLAGE

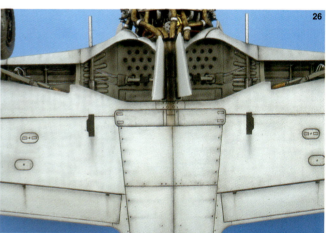

23, 24, 25. The weathering on the lower surfaces is done with airbrushed Tamiya paints, and then brush applied Prince August acrylics and oils. Note the colour of the ailerons that differs slightly from the rest of the wing surfaces. The panel outlines, here clearly marked, can be changed to suit the modeller's taste.
26. The undercarriage wells and the engine are full of detail and really make the model stand out.

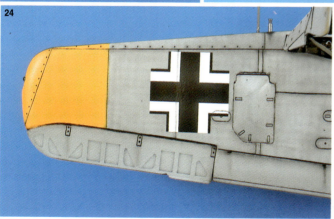

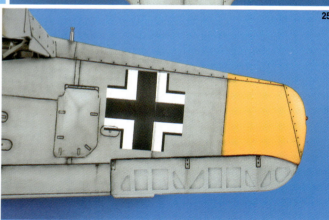

27, 28. The aircraft's tail seen from different angles giving us a good view of the yellow RLM 04 badges and tactical markings on the fuselage and at the base of the rudder.
29, 30, 31. The BMW 801 D-2 viewed from various angles with its twin star 14 cylinders cooled by a ten bladed fan. Equipped with a variable pitch VDM metal 3.30 m diameter propeller. Power was 1700 hp at take off up to 1440 hp at 5800m.
32. The flat canopy of the Fw 190 was equipped with an antenna tensioning system that kept it taut even when the canopy was completely open unlike the previous bulged canopy. In fact, two types of flat canopy were made. The first series was characterized by an almost continuous profile from the cowling to the forward post. The second, more common canopy juts out more at the top.
33. The starboard wing with the small red undercarriage down indicator. As well as the wing root mounted cannon, this Fw 190A-5 had two others, sometimes mounted optionally on this version that fired wide of the propeller.

Mediterranean but this did not stop them from being remarkably well suited to the East, especially in the summer in the vast, sun-baked steppe.

Painting the model

The camouflage is then airbrushed on but this time using acrylics whilst the wash applied later, is done with Titan oil based paint.

The hatches, cowling parts and other mobile parts are painted separately making sure that the shade of camouflage matches that of the rest of the model.

The finished model. This sort of project requires the modeller to be well organized as the detailing, assembly and painting are done simultaneously going back and forth between these jobs.

THE FW 190 F-8, DESIGNED for ground attack was the second from last of the F versions with 385 planes produced by Arado and Dornier. The Focke Wulf design team launched this specific series in April 1943 in order to harmonize all the ground attack variants (*schlachtflugzeug* or attack aircraft). The G version designated the Jabo-rei (*jagdbomber mit vergroesserter reichweite*, meaning long range fighter bomber). The F-8, which derived from the F-3, was built in 1944 in parallel to the A-8 series. It was armed with two engine mounted heavy MG 131, 13 mm machine guns and two MG 151, 20 mm cannon as well as being capable of carrying bombs, firing rockets, anti tank rockets, torpedoes or torpedo bombs (F-8/U-1 and U-2 versions.) The Fw 190 F-8/R-1 which we are particularly interested in here was equipped with a central ETC 501 rack and two under wing ETC 50/71 racks on which 300 litre tanks were fitted or bombs, from small 50 kg to 1,000kg for the heaviest as well as the more standard 250 kg and 500 kg.

This version does not require any important modifications like the two previous ones. We will simply improve the original kit with different commercially available additions like the AB 250 bomb by Verlinden. We will concentrate our efforts on two crucial points - the building, and the painting of the curious winter camouflage.

We will try and show every stage with the help of as many photos as possible. So much for the introduction; now we can get down to work.

Winter camouflage in detail

Achieving a winter camouflage scheme is always beset with doubts and problems. The main element, that of applying a white paint, is never easy.

(continued on page 56)

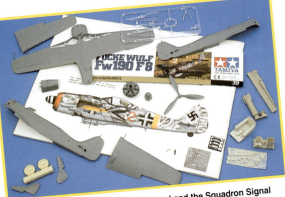
The Tamiya kit, the accessories used and the Squadron Signal (Number 1019) that deals with the Fw 190....

1. **The undercarriage wells are detailed with copper wire, pieces of plastic from electrical wiring, Evergreen sections and four Eduard reinforcement panels.**

2. **The lower part of the undercarriage doors was often removed on Fw 190s based on muddy or snowy airfields in order to prevent any blockages when retracting the undercarriage. The lower doors are therefore cut off using a blade as well as the small rectangular part below. The edges are then smoothed off with sand paper.**

3. **The joins between the bay walls and the internal wing profile are eliminated using pieces of Evergreen section.**

4, 5, 6. **We make the tail wheel using that of the Dragon kit (left on the first photo). It is more realistic than the one provided by Tamiya. We then separate the wheel from the fork so that we can show it turned. For this we use a hypodermic needle pushed into the shock absorber. We finish it off with a Dragon crutch, which is more realistic than the Eduard part, which is too flat.**

7. **The different undercarriage parts await painting. We have added cables and replaced the wheels with those of True Details.**

Unlike the previous chapter, the engine will hardly need any detailing. The Tamiya engine will be fine as it is because once the propeller is added it will not be visible. We will therefore paint it black and dry brush it with enamel aluminium. Note that the internal sides of the engine cowling, undercarriage wells, doors, shafts and torque links are painted in RLM 02.

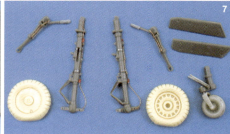

FOCKE WULF 190F-8/R1
‹ PANZERBLITZEN IN ACTION ›

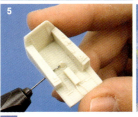
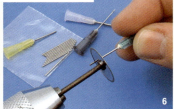

THE COCKPIT

1. We use a CMK resin set to improve the interior, completed by a few Eduard photo etched parts.

2. A selection of commercially available cockpit tubs, second from left is the CMK part that we chose and improved.

3, 4. We gently sand the joins of the resin parts such as the instrument panel, rear shelf or even the armour. Placing the parts on a flat surface, gently rub using sand paper in order to get rid of any surplus mould and make the parts as slender as possible. We advise you to moisten the sand paper as this will help to control the sanding when eliminating any unwanted resin.

5, 6, 7, 8. We next drill a 0.7 mm hole in the right side of the cockpit tub in order to install a 0.7 mm diameter metal rod representing the joystick control linkage. On the left side of the tub we need to add a throttle lever (2 mm long, 1.1 mm diameter), which we will finish off with Eduard and Reheat parts. The rudder bar supports are made from two joined needle sections, 0.9 and 1.2 mm in diameter.

9. Out of the seats shown here, the Cutting Edge seat (centre) is the most detailed and complete. The Tamiya part on the left is also fairly true to the original but the CMK seat (second from left) back is too wide.

10, 11, 12. The Cutting Edge seat with its seat belt completed with CMK photo etched parts, made with the help of a small vice and ruler.

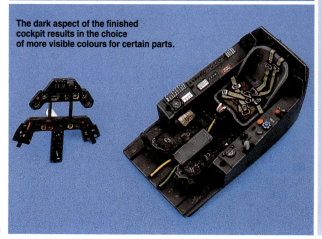

The dark aspect of the finished cockpit results in the choice of more visible colours for certain parts.

13

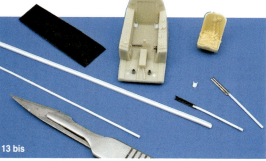
13 bis

14

13, 13bis and 14. We next cement the metallic parts on a rectangular Evergreen rod (ref 121) and make the base with a U section (ref 261) of which we round off the sides. We next cement the parts using cyanoacrylate applied to the grooves with a piece of copper wire.

15. The area between the two supports is sanded next using sandpaper fixed to a flat piece of wood.

16, 17, 18, 19, 20. The joystick is cut then its base is drilled so that we can add the control lever and missing cables.

15

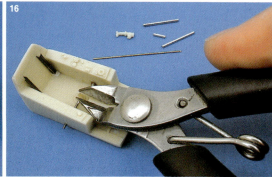
16

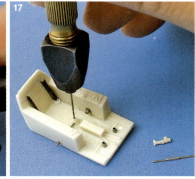
17

18

19

20

21

21, 22, 23, 24. Even though CMK has included the clock on the instrument panel we can refine the detail further still. First a hole is drilled to make way for a Reheat dial. A housing is then cut from Waldron sheeting using a punch. This is then cemented to a plastic rod and inserted from underneath. (leaving a slight difference in level). Looking at numerous photos of the instrument panel, the Waldron set (ref 4803, set number 9) is a perfect reconstitution of the instruments.

25. A new selection of parts, this time for the rear of the cockpit.

26. The CMK part that was chosen will be cemented to the pre-sanded original. Note that the pilot's luggage compartment is separate.

27 and 27bis. The CMK rear armour is placed on the support characteristic of the Fw 190 bulged canopy.

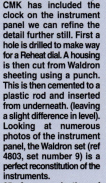

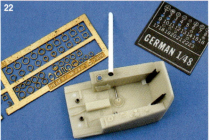
22

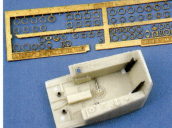
23

24

25

26

27

27 bis

51

28, 28 bis, 29, and 30. Of all the available rudder pedals we chose the Waldron. They are mounted on a scratch built rod (brass rods 0.5 mm diameter, metal and plastic tubes) with the photo etched straps and fitted in the tub.

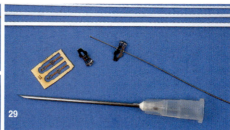
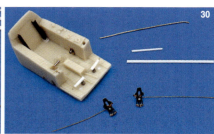

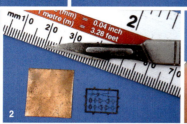
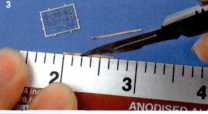

1. The missing levers are made using Reheat (ref RH 45) parts and Evergreen 0.9 mm diameter rods.

2, 3, 4, 5 and 6. The seat belt is made from tin sheet, cut out with a blade and ruler. It will be finished off with Waldron (ref 4806) buckles.

We next attach the belts to two small copper hooks at the rear of the cockpit tub and finish off by adding the two small photo etched Eduard lateral sheets to the tub.

7. The improvements to the cockpit sides are achieved with the help of Evergreen sections (ref 261, 100 and 101)

8, 9, 10. We begin by getting rid of all the parts that jut out. We sand down the surfaces and file and sand paper down the walls. The difference before and after this process is obvious. (cf right hand wall)

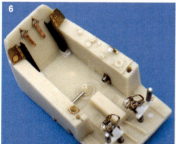
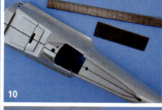

11. This having been done, we cement the Evergreen canopy rail sections.

12. The canopy crank handle is made from a photo etched Dragon disk, a thinned down Evergreen section (ref 101), a CMK knob and a Reheat part.

13. A photo showing the improved fuselage and the original.

14. 0.2 mm rivets are drilled along the sections added to the internal walls.

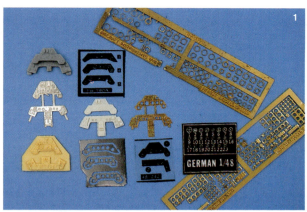

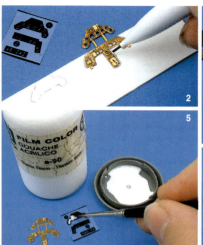

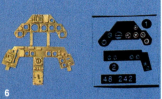

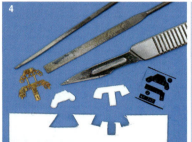

1. The instrument panel can be either chosen from a wide range of manufacturers or scratch built. In an anti-clockwise direction we have the Tamiya part, followed by Dragon, Verlinden, Eduard and CMK in the centre.

2, 3 and 4. We decided to choose the Eduard part, but lined with 0.5 mm plastic sheet. This new part is cut out using the outline of the photo etched part. It is then rubbed down with a file and sandpaper.

5, 6 and 7. White acrylic paint is applied to the reverse of the film allowing for a better representation of the dials and instruments.

8. Kristal Klear is used to fix the film onto the plastic sheet copy of the instrument panel.

9. The instrument panel is extremely well cemented thanks to holes drilled into the lateral control panels.

10. The instrument panel is painted in RLM 66 and the instruments on the lower panel are painted in black.

11. Various dials are painted in blue, yellow, red and light grey. It is preferable to consult good documents.

12. After painting, the control panel is cemented onto the film using Kristal Klear. This allows you to rectify any mistakes. It is then fixed to the plastic sheet support.

Below. The instrumental panel made by Waldron appears really the best when you consult good documents.

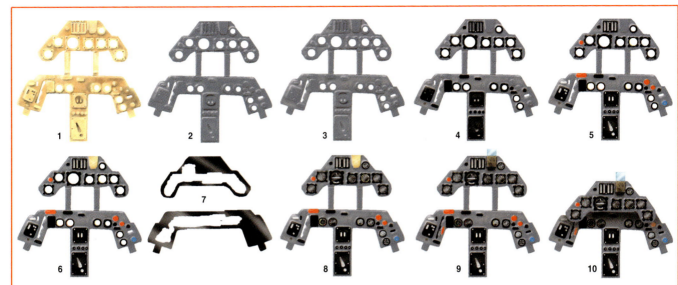

SUMMARY OF THE DIFFERENT STAGES OF THE INSTRUMENT PANEL PAINTING

1. Photo etched Eduard part before painting
2. Airbrushing in RLM 66
3. Highlighting in order to achieve volume.
4. Brush painting in black of the different dials and instruments.
5. Adding details using various colours.
6. Cleaning the paint around the Revi 16 gun sight emplacement.
7. Applying white paint to the reverse of the film to bring out the instrument details.
8. Using Kristal Klear to fix the film to the rear of the photo etched **part**
9. Fitting the Adeco Revi 16 sights.
10. Folding the photo etched flaps to create the shape of the instrument panel.

(photos by Eduard Soler)

13. All the parts are ready for painting.
14. The smaller parts are fixed to various supports that will facilitate airbrushing.
15. The first coat of RLM 66 to the inside of the cockpit, which is partially marked off with adhesive tape.
16, 17. The second coat (over a larger area) using a very diluted and lightened paint. This will give an impression of volume.
18. We next add a very light and limited wash using diluted black acrylic Prince August.

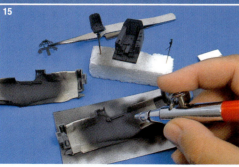

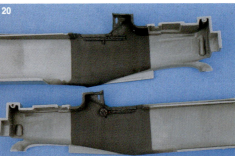

19. All the detailing (seat belt, buttons, panels) is brush painted in different Prince August acrylics.

20. The sides are also brush painted and completely finished before installing the seat pan. Note that the slide rails are highlighted using a little light paint.

21, 22. The finished cockpit is placed temporarily next to the fuselage halves. Because of its fragility the Revi 16 gun sight will be added later.

23. The cockpit has been fitted and you can see the gun sight mount on the instrument panel hood. The gun sight screen will be fitted at the last moment when all painting and assembly is finished.

PAINTING THE INTERIOR

— **Cockpit interior**
RLM 60 (lightened Tamiya XF-63).
The instrument panel is darker and the seat lighter.
— **Seat belt.** Prince August 987

— **Other details.**
Black, light grey, red, blue, yellow, green,
Prince August white, Tamiya X-11 (enamel range)

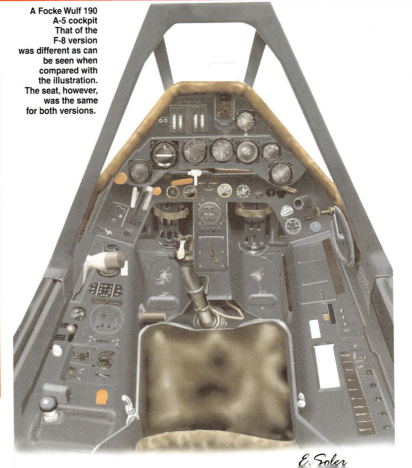

A Focke Wulf 190 A-5 cockpit That of the F-8 version was different as can be seen when compared with the illustration. The seat, however, was the same for both versions.

E. Soler

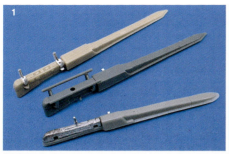 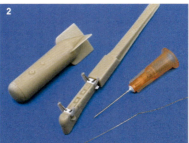 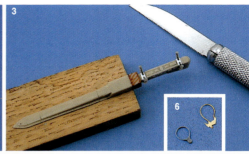

1. The ETC 501 bomb rack as shown from top to bottom by Dragon, Tamiya and Verlinden. The latter, despite being a little too long and in need of improvement, is the best finished.

2. The rack fixations are made with a very thin hypodermic needle, a brass 0.5 mm diameter rod and Evergreen sections.

3. After consulting the Aerodetail fact sheet on the Fw 190 A we see that the rack is too long and we saw off the area marked in red.

4. Fixing the rack between the undercarriage wells, checking that it is well aligned with the belly.

5. The AB 250 fragmentation bomb and its rack, both by Verlinden. It is possible to combine several types of armament, for example four SC 50 bombs loaded on their ETC 71 racks (these are perfectly made by Tamiya and do not need replacing with Verlinden parts.) This can be completed by a drop tank loaded onto an ETC 501 rack.

6 and 7. The FuG 16ZY radio transmitter was equipped with two types of antenna, a classic wire and another hoop shaped. For our model we can use the Tamiya part by thinning it down as much as possible or make our own from a piece of Evergreen. We can see that photo etching does not succeed in reproducing this sort of part very well. The Gonio, top left, is too flat and the hoop is incorrectly positioned.

(continued from page 48)

The challenge, if we can call it that, for this version is twofold as the camouflage is composed of white stripes painted onto the standard Luftwaffe RLM 74 and 75 scheme. We have to paint a light colour onto a dark base.

The plane we chose is a Focke Wulf Fw 190 F-8/R1 of the 1./ SG2 based in a snowbound and frozen Hungary, January 1945. The German Army at this point was forced to retreat progressively under the pressure of the Soviet Army.

This plane was mostly used in a ground attack role like the aging Stuka, and at the same time was able to defend itself relatively well if intercepted by Russian fighters.

The visual documents showing this plane are not of much help when it comes to dealing with the camouflage. On a different subject we can see that the lower

1, 2 and 3. We begin by washing the kit in soapy water then mask the transparent areas of the canopy. For this we use the masking from the Eduard XF 009 sheet and Maskol.

4. The edge of the cockpit needs to be masked off and protected with paper so that the upper area can be painted problem free.

5. We then airbrush on a first coat of dark grey RLM 66 which is obtained by mixing Tamiya paints, not forgetting the internal canopy parts.

6. A veil of very diluted RLM 66 mixed with white adds the first highlights and volume.

part of the undercarriage doors have been removed. This helped the plane taxi and take off from weather-affected airfields and ensure that the undercarriage retracted correctly.

It generally carried an AB 250 fragmentation bomb and a wing mounted reconnaissance camera.

The latter being, however, an accessory and largely dependant on the type of mission, we have not included it, a choice made all the easier by the fact that no relevant information was found.

Concerning the specificity of the camouflage and markings and in spite of the various interpretations by illustrators, we note that the wing tips were not in RLM 04 yellow but had a yellow V (a tactical recognition marking) on the left side only.

Note the camouflage and panels are airbrushed using Prince August acrylics. All colour toning is done using the Federal Standard colour chart.

A lack of paint

The fuselage band was also narrower than the Luftwaffe standards dictated and positioned closer to the black crosses.

The cowling ring was painted in red RLM 23 with the lower panels in yellow RLM 04. This Fw 190 F-8/R1 was equipped with a bulged canopy and reinforced armour plate characteristic of later models.

The Fw 190s equipped with this type of canopy did not have the roller assembly for the radio antenna whereas the flat canopy did. This meant that the antenna cable slackened when the canopy was opened allowing part of it to fall against the fuselage.

Two explanations are possible when it comes to the washable white stripes, an unknown scheme up to that point and different from the whitewashes that suited the operational environment.

The first, a specific need considering the environment in which the plane operated or, more mundanely, a lack of paint. The latter is backed up (with all reserve) by the fact that the cowling is almost completely plain.

The white paint chosen is from Tamiya XF-2 acrylic. It is filtered through nylon tights to catch any lumps and is strongly diluted with 96° alcohol to make it more fluid and more easily airbrushed.

Several coats were needed to cover the original grey camouflage with which we will start the long painting process.

Once again we invite you to follow the whole process of this kit's construction through the numerous photos. These will show the smallest details and the problems met. Building and finishing secrets will be revealed, especially concerning the riveting which will keep us busy for quite a while.

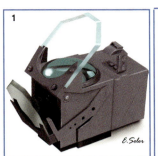
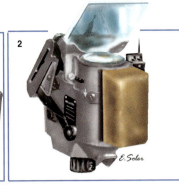

1. Revi 16 B gun sight

2. Revi 12 D gun sight with its characteristic anti shock cushioning.

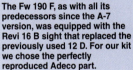

3. The gun sight having been fitted we finish off by acrylic painting the small details.

4 and 5. The edges of the rear compartment hatch, the hatch rivets and finally, cementing the windscreen with Kristal Klear.

The Fw 190 F, as with all its predecessors since the A-7 version, was equipped with the Revi 16 B sight that replaced the previously used 12 D. For our kit we chose the perfectly reproduced Adeco part.

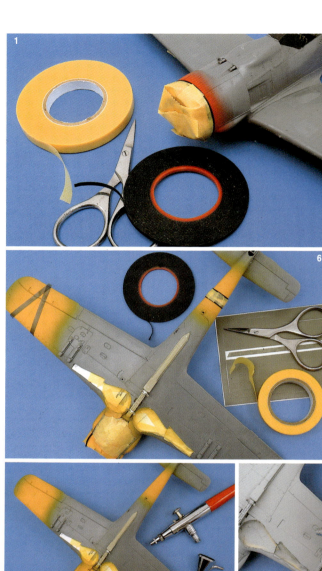

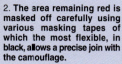
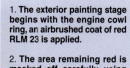

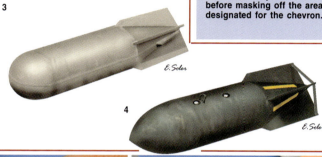

1. The exterior painting stage begins with the engine cowl ring, an airbrushed coat of red RLM 23 is applied.

2. The area remaining red is masked off carefully using various masking tapes of which the most flexible, in black, allows a precise join with the camouflage.

3 and 4. The SC 500 bomb with its four tail fins and the AB 500 with its rounded nose. Both weigh 500 kg. The Fw 190 F we are building carried an AB 250 fragmentation bomb weighing 250 kg.

5. The masking process continues using thin card and Tamiya adhesive tape to the undercarriage wells and cockpit.

6. We next add yellow RLM 04 before masking off the area designated for the chevron.

7 to 13. A series of pictures showing the different stages of the camouflage. The belly and flanks are painted first in RLM 76, followed by the upper surfaces and flanks in grey RLM 74/75.
At this point we add that these stages are done in a continuous manner without using masking, taped or otherwise. This stage of the painting is finished with the fairly dense mottling, composed of equal daubs of RLM 74/75.
From a technical point of view it is necessary to dilute the paints well to avoid any unwanted drips. It is also necessary to trace the bands of RLM 74 beforehand. It can then be applied, airbrushing from the outside to the inside of the areas to be painted.

14, 15 and 16. We next use Tamiya XF-2 acrylic for the RLM 21 winter camouflage. This is a particularly delicate stage due to the colour's lack of suppleness. Because of this it is essential to dilute it then filter it through nylon tights so that any lumps are eliminated. Being well diluted, the white paint's covering properties are weakened and several coats are necessary.

ADDING AND PAINTING THE DECALS

17, 18. The whitewash finished, the areas destined for the wing crosses are touched up with a little RLM 74 grey around the approximate contours.
19 and 20. Once these areas are painted the Eduard stencils (ref XF 525) for the white crosses are added.
21 to 23. The same applies for the fuselage crosses. The stencil is placed and its edges protected with adhesive tape. The black paint is applied and the stencils removed so that a few white finishing touches can be applied to simulate the overlapping of the white camouflage.
24. The exhaust panels are first painted in black.

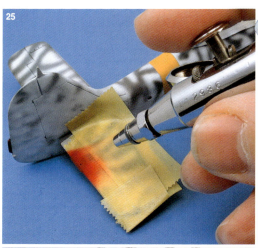

25. The aileron and rudder trim tabs are always red on Fw 190s. The masking is applied over a large area to protect the wings from any eventual paint splash.

26, 27. The first stage of the painting is now finished. The belly and upper areas of the plane are new and show no signs of wear and tear. All areas destined for decals and markings now receive a coat of gloss varnish.

28. Applying the decals is done with the usual softening products, here the equivalents of Micro Sol and Micro Set by Microscale, these are readily available from specialist shops.
It is important to cut away as much of the decals' transparent film as possible using a new blade and also avoid applying too much gloss varnish.

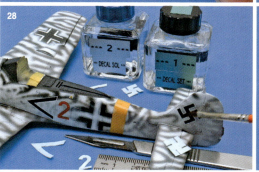

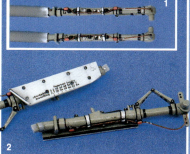

1. The undercarriage legs are painted in RLM 02 with the details brush-painted with acrylics. The markings are gloss varnished *EagleCals* decals.
2. The legs with their high hatches, the door panels having been fitted.
3. The undercarriage had a characteristic splay, inclined towards the interior (in 1/48th scale, the distance from one wheel to another, from the middle of each one, must be 7.3 cm, respecting the 3.5 m of the full sized plane) whilst being slightly inclined forward when seen from the side. The small pieces of white card protect the chromed shock absorbers when the coat of matt varnish is applied.
4. The propeller blades are painted in RLM 70 green/black with an airbrush applied lightened, weathered effect.

Left.
The Pitot tube is made from two hollow Mimeca metal rods inserted into each other. The lower surfaces of the right wing with the same remarkable brush painted riveting so long to achieve. The panel lines are airbrushed.

1. We begin by diluting a little grey oil based paint with turps. This is then applied along the panel lines.

2. We next apply to the cowling area an oil blend with a well diluted white base and RLM 74/75 greys. Apply the mixture from top to bottom.

3, 4. To enhance the panel lines we chose the technique of masking and airbrushing which gives a subtle and blended effect. Each line is masked off with Tamiya tape then delicately airbrushed with a mix of grey Tamiya XF- 24 (10%) and thinner (90%). Keep as close as possible to the masking tape.

5. By painting over onto the masking tape you can control the density of the paint, the airflow and the quality of the discharged paint. The desired effect should be perfectly consistent, fairly light and soft. To achieve this the pressure should be fairly weak, under 1 bar if possible.

6. The belly having been thus treated we move on to the upper surfaces varying the shades depending on whether the area is light or dark, when the line goes through the wing cross for example.

7, 8. The tail plane maintenance hatch is done in the same way.

9 to 11. The canvassing is prepared in order to highlight the contours of the internal struts. This long process tries to obtain contrast thanks to the bands applied to the protruding areas. The following airbrushing should be very precise in order to cover just the outlines of the ribs. When finished these should appear lighter.

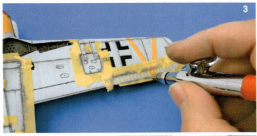

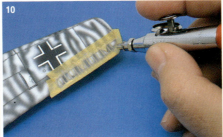

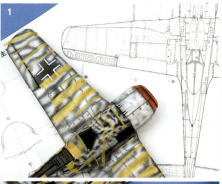

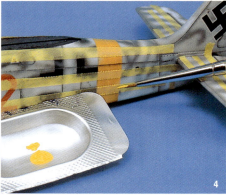
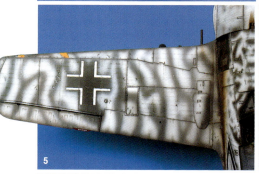

The rivets

Modellers, each time they begin a new project, tend to want to do better than the last time, setting themselves real challenges.

One of these challenges is representing the rivets, something that is talked about more and more in specialist magazines and on the internet.

Depending on the plane and its condition, the task can be relatively complicated such as with the Fw 190 and its 15000 fuselage and wing rivets!

The correct rounded shape is difficult to achieve and must also respect the scale.

It is important to strike the right balance in shading and highlighting that expresses the manufacturing methods or the way that the metal reacts under different structural stresses. As far as painting the rivets is concerned, a lighter tone is used than that of the paint in the structural lines.

In this way, we can, if we take into account these factors and our own technical limitations, vary the effect produced and improve the appearance of the model which will eventually look very different from the usual representations that we have been used to seeing for decades.

Let's now see what this means in practice for the Fw 190F-8/R-1.

1. It is absolutely essential to have a good scale drawing when tackling the rivets. Patience is also essential!

2. We begin by placing thin strips of adhesive tape along the fuselage and wings where the rivets are supposed to be. These strips will help guide the work and help in the painting.

3. This stage is carried out with a thin brush and Prince August acrylics of a lighter colour than the camouflage.

4. For the yellow fuselage band the rivets are shown as small dots.

5 and 6. The work continues and is finished with the airbrush, always following the strips but trying, with the appropriate colours to lessen the hardness and clarity of the brush strokes.

INTERIOR AND CAMOUFLAGE COLOURS

— **COCKPIT**
RLM 66 (Prince August 994)
openings and wells RLM 02 (Prince August 886)

— **DETAILS.**
Black, red, yellow,
different browns and greys
plus some touches of enamel silver,
Tamiya XF-10, XF-11 and XF-16.

— **ENGINE.**
Prince August acrylics, Tamiya Enamel
black acrylic wash, silver dry brushing in Tamiya XF-11.

— **UNDERCARRIAGE AND TYRES.**
undercarriage and internal hatches RLM 02
tyres, grey black (Prince August 862)

— **CAMOUFLAGE**
Lower surfaces. RLM 76
Upper surfaces. RLM 25/79/80
Tactical markings. RLM 04
Propeller spinner. Black and white spiral
Propeller blades. RLM 70

— **PANEL HIGHLIGHTING.**
Dark grey XF-24
and Deck Tan XF-55 varnish

— **APPLYING THE DECALS.**
Gloss acrylic Micro Gloss

— **FINISHING TOUCHES**
Synthetic Marabu varnish, matt or satin.

1, 2, 3 and 4. The best way of fitting the canopy is to place a brass rod in the faired armour reinforcement and cement it with cyanoacrylate in a hole made a third of the way along the central groove.
Use as little cement as possible as this will reduce the risk of any discolouration caused by the vapours. White glue can also be used but it is not as strong.

A and B. The airbrushed Tamiya acrylic must be very diluted. We begin with a veil of sand XF-55. The effect we are after takes into account the following parameters. For areas subjected to the hottest gases, the tones will be dark on a light base and light on a dark base (as seen in the openings) the tone lightens as you move farther away from the exhausts.

C. The grooved machine gun channels are more or less painted in the same way as the exhausts. The red armoured cowl ring shows various marks and spills caused by the engine. These are applied with a fine brush.

D. The lower cowling with its brush applied rivets and well marked panel lines. The gas evacuation tubes also stand out.

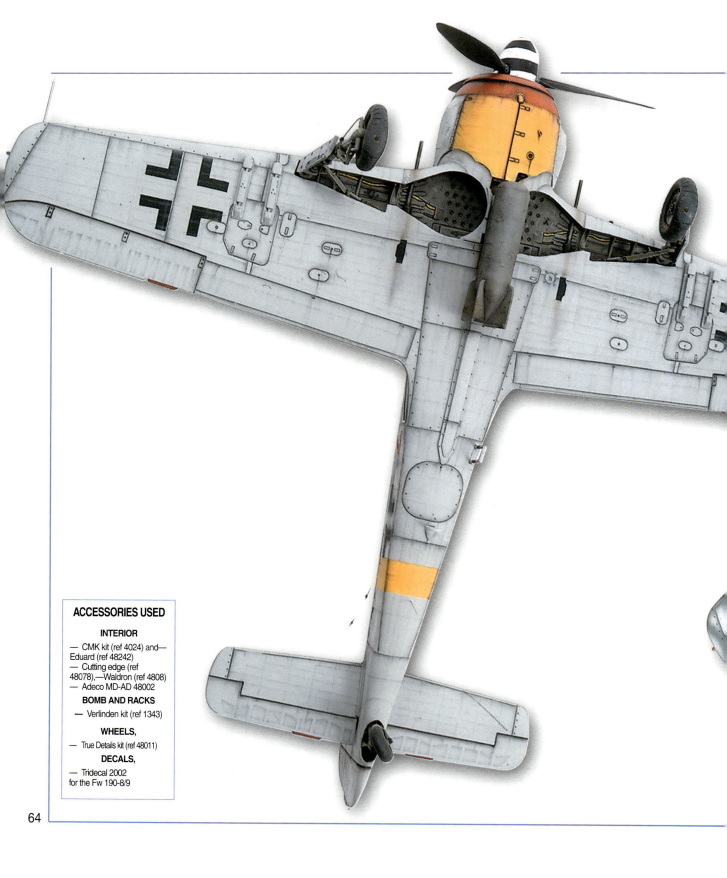

ACCESSORIES USED

INTERIOR
— CMK kit (ref 4024) and—Eduard (ref 48242)
— Cutting edge (ref 48078),—Waldron (ref 4808)
— Adeco MD-AD 48002

BOMB AND RACKS
— Verlinden kit (ref 1343)

WHEELS,
— True Details kit (ref 48011)

DECALS,
— Tridecal 2002 for the Fw 190-8/9

A very fluid brown patina is applied to the camouflage tones of the wings, especially along the trailing edges.

FOCKE WULF 190F-8/R1
« PANZERBLITZEN IN ACTION »

From left to right.
The stripes are neither too bright nor too soft, just enough to keep to scale and the way in which it was probably applied.

The left hand lower surfaces with the lines of rivets on a RLM 76 base, the strange yellow chevron, hinges and the marks behind the cannons and shell casing ejectors.

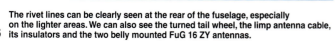

The rivet lines can be clearly seen at the rear of the fuselage, especially on the lighter areas. We can also see the turned tail wheel, the limp antenna cable, its insulators and the two belly mounted FuG 16 ZY antennas.

Darkening the panel lines by masking is a technique that gives a very pleasing result but it is not the only technique.

The bomb is touched up here and there with a brown paint to show oxidation. Other lighter marks are added crosswise whilst the base of the tail fins are highlighted in black. The undercarriage cavities also receive a dark wash.

The finished model after a long painting session.

Fw 190A-3 from the 9./JG 2, France, winter 1942-43.

A CAMOUFLAGE SCHEME
Step by Step

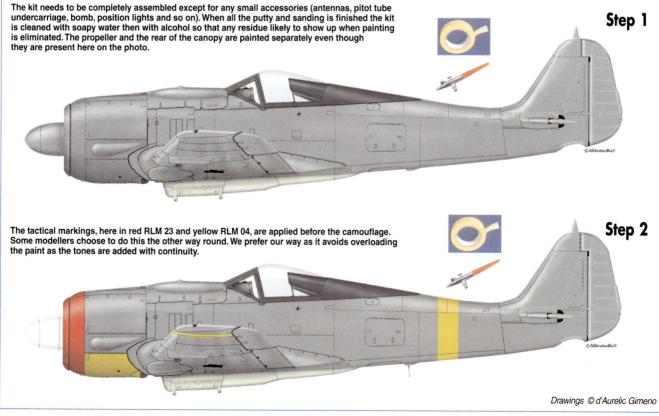

Step 1 — The kit needs to be completely assembled except for any small accessories (antennas, pitot tube undercarriage, bomb, position lights and so on). When all the putty and sanding is finished the kit is cleaned with soapy water then with alcohol so that any residue likely to show up when painting is eliminated. The propeller and the rear of the canopy are painted separately even though they are present here on the photo.

Step 2 — The tactical markings, here in red RLM 23 and yellow RLM 04, are applied before the camouflage. Some modellers choose to do this the other way round. We prefer our way as it avoids overloading the paint as the tones are added with continuity.

Drawings © d'Aurelic Gimeno

Step 3

The markings are masked off before applying light grey blue RLM 76 on the belly, flanks and tail plane.

Step 4

The main camouflage is the next stage. As always we begin with the lightest colour, grey RLM 75 applied freehand with an airbrush being careful to do fine lines. Unlike what the photo shows, the masking on the tactical bands and canopy should be left in place during the painting process.

Step 5

We carry on with the second grey, RLM 74. This is applied in the same way with the masking still in place.

Step 6

We finish off the camouflage by adding numerous small mottles. The model is now ready for its winter camouflage.

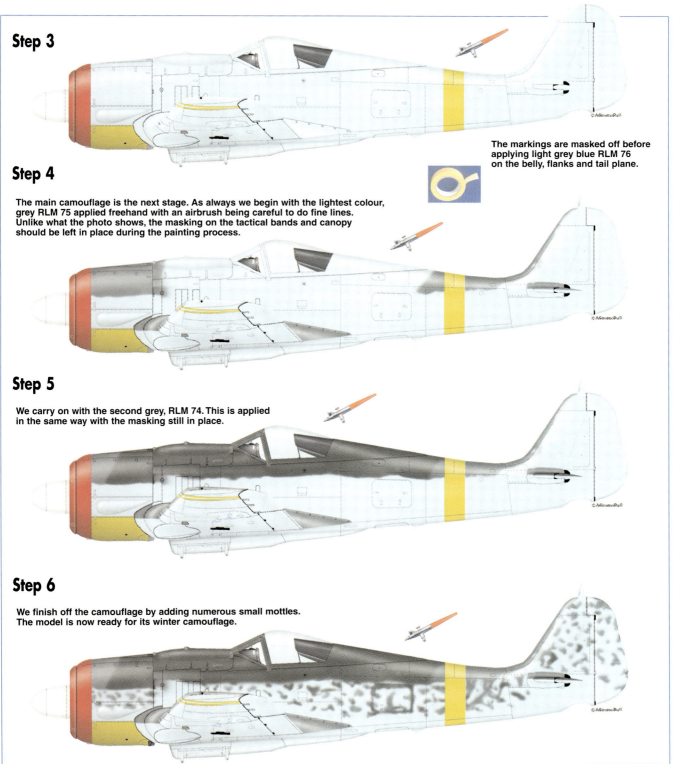

Step 7

The white acrylic paint used here will be filtered through nylon tights that will retain the heavier lumps. The paint must be very diluted and applied in several coats in order to cover the dark tones of the camouflage. It is applied with an airbrush from close up and at a low pressure (between 0.5 and 0.8 bar) thanks to a variable flow compressor. Note the continuous scheme on the cowling compared to the more marbled effect on the fuselage.

Step 8

In this stage the wing crosses are applied using Eduard stencils. Other decals are also added. These are placed on a coat of gloss varnish using softening products. We next add a few airbrushed touches of diluted white paint, overlapping a little onto the crosses and the fuselage band at the rear. We finish off by painting the areas around the exhausts.

Step 9

The main part of the painting process is finished off by applying a little dark grey oil wash along the structural lines. Exhaust marks are finally painted, all small parts cemented and any finishing touches applied. A coat of Marabu matt satin varnish will protect the kit's surfaces and concludes this stage.

Note that in general we used Tamiya acrylic paints for airbrushing, Prince August for brush painting and oil paints for weathering.

REICH FUSELAGE BANDS

JG 1 March 1944 to April 1945 and JG 300 until December 1944.

JG 2 from October 1944 or January 1945

JG 3 from June 1944 to May 1945

JG 4 late period (beginning 1945)

JG 5 late period (beginning 1945)

JG 6 late period (beginning 1945)

JG 11 from the beginning of 1944 (January or February?)

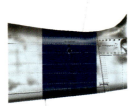

III./ JG 54 from the beginning of 1944 then to all of the Wing the following year

JG 300 late period (beginning 1945)

JG 301 late period (January 1945)

At the beginning of 1944 all planes belonging to the same Wing received coloured bands at the rear of the fuselage just in front of the tail plane. This was a measure destined to facilitate the identification of planes attached to the Reichverteidigung (defence of the Reich). There were six different colours for these bands (red, yellow, green, blue, black and white) and they were exactly 900 mm in width. This was, in fact, a normalization of what was already in use with JG 1 and JG 300 whose planes bore a red band.

The misfortunes of war meant that an increasing number of planes were allocated to defending the Reich. This led to a new standardisation instigated by the OKL (Oberkommando der Luftwaffe, the general staff of the Luftwaffe) that increased the number to 17 possible options in February 1945.

The band consists of one monochrome strip 900 mm in width, or two of 450 mm or finally three of 300 mm of two different colours only but always identical to the six initial ones.

This standardisation was brought into being after certain bomber interception units began using new colour combinations all of which were non-standard. Below we have the various colour bands used on the Fw 190A along with the dates, except for JG 26 for which we do not have sufficient documentation. (Note: This band was black and white according to our sources)

Note. The dates indicated correspond to the publication of the standardisation even though certain bands were used a long time beforehand. Towards the end of the war organization began to break down and it is highly probable that some bands were not used by all planes from the same wing.

THE NOSE AND THE COLOURS

The BMW radial engined Focke Wulf Fw 190A had the most colourful cowling of all the Luftwaffe's fighters in the Second World War.

Those of JG 1 were particularly spectacular, decorated from 1943 with a superb chequered pattern whose primary colour corresponded to the Wing (Staffel) it belonged to.

The cowlings were painted in other ways, black and white bands for example, that could coexist at the same time and in the same unit as planes whose noses were monochrome, either white or yellow.

These uniform noses were not, however, only specific to JG 1 as assault groups (Sturmgruppe) of JG 3 adopted red and black, common colours at the time.

Note that the lower panel was in principal painted as a means of rapid identification, yellow for the northern and central Russian Front sectors and white for its southern sector, Mediterranean and North Africa.

There were also some stranger schemes like those of JG 2s Fw 190 A-1 planes in the spring of 1942. These had eyes and an open mouth, or more even outrageous ones like Herman Graf's famous Fw 190 A-5 with its red flames and yellow borders: (Jagderganzungsgruppe Ost, Bussac, southern France).

Below is a selection of the most striking Fw 190 A cowlings with the exception of the JG 2 eagle heads that covered a larger area in order to cover the exhaust stains.

COWLINGS

1. 1./ JG 1, 1943
2. 2./ JG 1, 1943
3. 3./ JG 1, 1943
4. I./ JG 1, scheme used from the beginning of the war until mid 1944
5. 6./ JG 1, 1943
6. 1./ JG 1, 1943
7. IV./ (Sturmstaffel) JG 3, Fw 190A-8
8. IV./ JG 3. A possible interpretation, black being for the moment the most likely colour for the Fw 190 A-8/R aircraft of IV./ JG 3.
9. 6 (Sturmstaffel). JG/300, November 1944. Fw 190 A-8/R2, piloted by P. Lixfeld.
10. SG 2, Eastern Front, winter 1944-45, Fw 190 F-8
11. III./ JG, 1945 (Fw 190 A-8)
12. 8 (Sturmstaffel)./ JG 300, February 1945, Fw 190 A-8/R8 piloted by V. Heimann
13. Unknown unit Eastern Front Fw 190 F-8/R1.
14. I./ JG 301 (Fw 190 A-8).
15. Jagderganzungsgruppe Ost, France 1943. Fw 190 A-5 piloted by Major Hermann Graf.
16. II./ JG 1, Holland, Spring 1942. Fw 190 A-1.

A Focke Wulf Fw 190A-4 of Stab JG 54. Estonia, Summer 1943. We can see that the camouflage layout and colours on the wings are similar to that of the fuselage.

Fw 190s COWLINGS, 1941-45

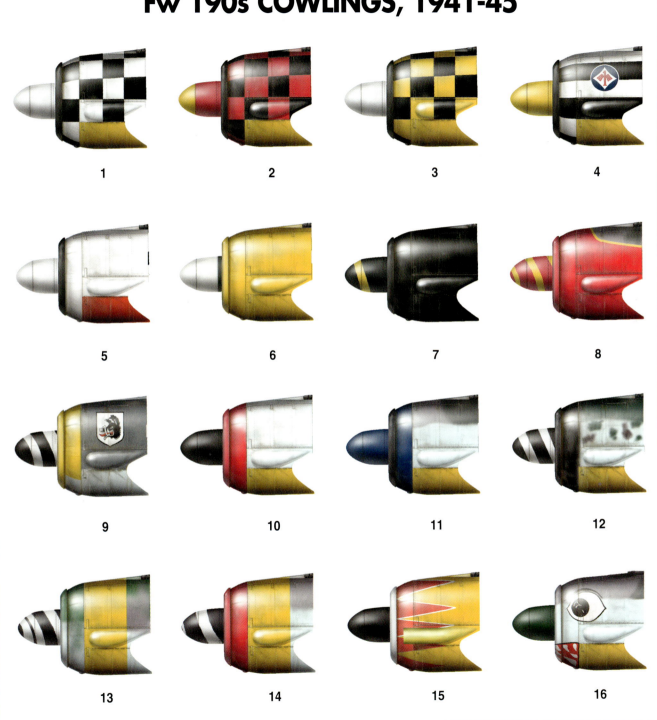

Fw 190 V-1, Reichlin test flight centre, Germany. Beginning of 1940. The plane has been modified and has lost its original ducted propeller spinner, which caused so many cooling problems. The code has also been changed and registered militarily as FO+LY. A number 01 is also present on the rudder and the tail plane no longer has the swastika that was initially present. The camouflage is standard, green RLM 70/71 applied in a splinter pattern with the belly painted in RLM 65. The last point of interest is the long one-piece canopy that does not have the metallic rear part present on production models.

An Fw 190A-3 from 8./JG 2 stationed in Théville, France 1942. For a while planes from this *Jagdgeschwader* were decorated with a beautiful eagle's head that helped hide the exhaust stains. This aircraft had a standard factory camouflage scheme with its upper surfaces in RLM 74/75 the belly in RLM 76 and identification markings in yellow RLM 04. Note the undulating line separating the fuselage colours and the way the crosses have been touched up to make them less visible.

An Fw 190A-3 from III./JG 2, landed by *Oberleutnant* Armin Faber on the Pembrey airfield in Great Britain, 23rd June, 1942 at 19.36 precisely. Faber made a navigational error after a dogfight with Polish Spitfires from the Exeter Wing and landed on the British airfield presenting the Royal Air Force, for the first time, with an intact Fw 190A-3. It was given a new serial number, MP 499 and was rigorously tested at the Royal Aircraft Establishment (RAE) against a Spitfire Mk. IX, a Typhoon and a Griffon engined Spitfire. The tests continued, this time against American aircraft and it finished its flying career on January 2nd 1943. The camouflage is RLM 74/75/76 with a mixed mottling, yellow markings on the tail fin and lower cowling as well as a curious white, black-bordered band around the tail.

Fw 190A-3 belonging to I./JG 5 . Herdla, Norway during 1943. Standard RLM 74/75/76 grey camouflage with yellow cowling underside. The large white 1 on the fuselage is unusual.

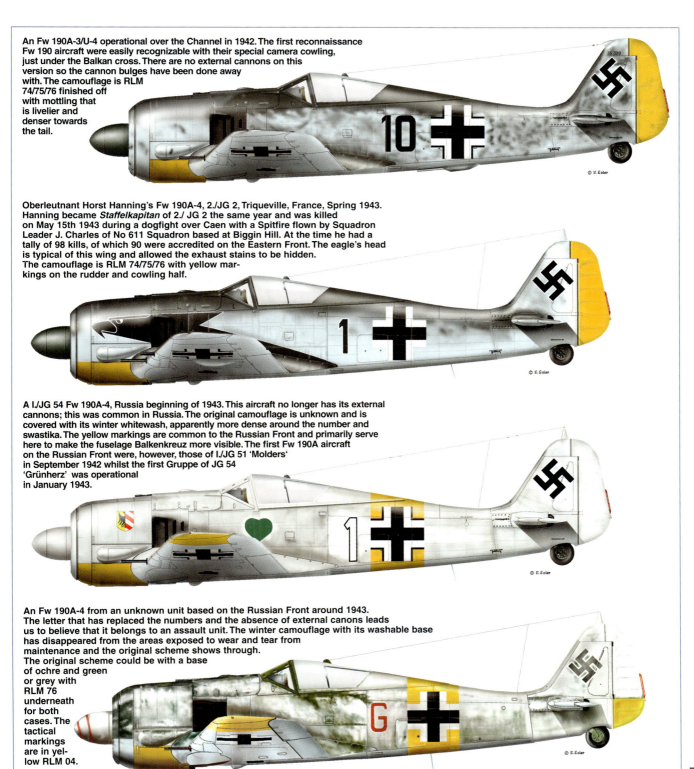

An Fw 190A-3/U-4 operational over the Channel in 1942. The first reconnaissance Fw 190 aircraft were easily recognizable with their special camera cowling, just under the Balkan cross. There are no external cannons on this version so the cannon bulges have been done away with. The camouflage is RLM 74/75/76 finished off with mottling that is livelier and denser towards the tail.

Oberleutnant Horst Hanning's Fw 190A-4, 2./JG 2, Triqueville, France, Spring 1943. Hanning became *Staffelkapitan* of 2./ JG 2 the same year and was killed on May 15th 1943 during a dogfight over Caen with a Spitfire flown by Squadron Leader J. Charles of No 611 Squadron based at Biggin Hill. At the time he had a tally of 98 kills, of which 90 were accredited on the Eastern Front. The eagle's head is typical of this wing and allowed the exhaust stains to be hidden. The camouflage is RLM 74/75/76 with yellow markings on the rudder and cowling half.

A I./JG 54 Fw 190A-4, Russia beginning of 1943. This aircraft no longer has its external cannons; this was common in Russia. The original camouflage is unknown and is covered with its winter whitewash, apparently more dense around the number and swastika. The yellow markings are common to the Russian Front and primarily serve here to make the fuselage Balkenkreuz more visible. The first Fw 190A aircraft on the Russian Front were, however, those of I./JG 51 'Molders' in September 1942 whilst the first Gruppe of JG 54 'Grünherz' was operational in January 1943.

An Fw 190A-4 from an unknown unit based on the Russian Front around 1943. The letter that has replaced the numbers and the absence of external canons leads us to believe that it belongs to an assault unit. The winter camouflage with its washable base has disappeared from the areas exposed to wear and tear from maintenance and the original scheme shows through. The original scheme could be with a base of ochre and green or grey with RLM 76 underneath for both cases. The tactical markings are in yellow RLM 04.

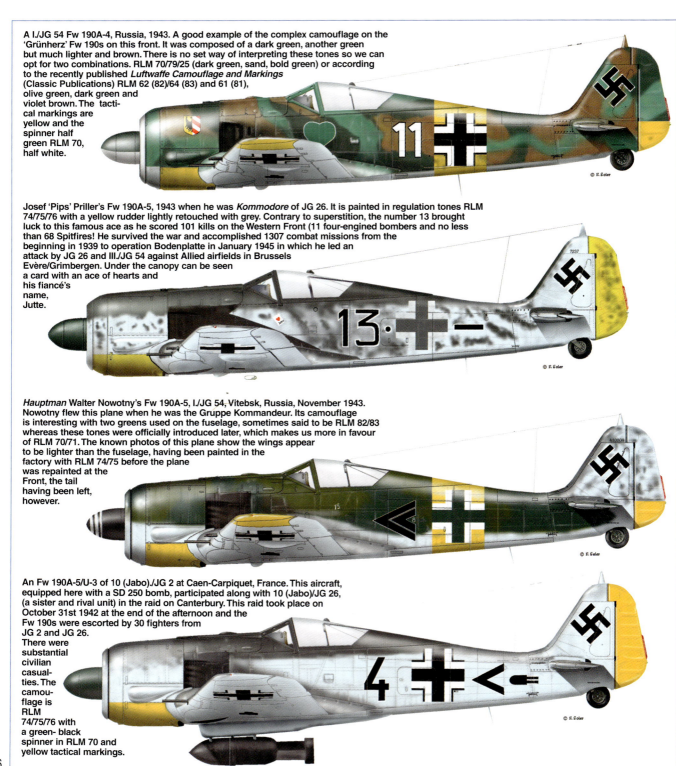

A I./JG 54 Fw 190A-4, Russia, 1943. A good example of the complex camouflage on the 'Grünherz' Fw 190s on this front. It was composed of a dark green, another green but much lighter and brown. There is no set way of interpreting these tones so we can opt for two combinations. RLM 70/79/25 (dark green, sand, bold green) or according to the recently published *Luftwaffe Camouflage and Markings* (Classic Publications) RLM 62 (82)/64 (83) and 61 (81), olive green, dark green and violet brown. The tactical markings are yellow and the spinner half green RLM 70, half white.

Josef 'Pips' Priller's Fw 190A-5, 1943 when he was *Kommodore* of JG 26. It is painted in regulation tones RLM 74/75/76 with a yellow rudder lightly retouched with grey. Contrary to superstition, the number 13 brought luck to this famous ace as he scored 101 kills on the Western Front (11 four-engined bombers and no less than 68 Spitfires! He survived the war and accomplished 1307 combat missions from the beginning in 1939 to operation Bodenplatte in January 1945 in which he led an attack by JG 26 and III./JG 54 against Allied airfields in Brussels Evère/Grimbergen. Under the canopy can be seen a card with an ace of hearts and his fiancé's name, Jutte.

Hauptman Walter Nowotny's Fw 190A-5, I./JG 54, Vitebsk, Russia, November 1943. Nowotny flew this plane when he was the Gruppe Kommandeur. Its camouflage is interesting with two greens used on the fuselage, sometimes said to be RLM 82/83 whereas these tones were officially introduced later, which makes us more in favour of RLM 70/71. The known photos of this plane show the wings appear to be lighter than the fuselage, having been painted in the factory with RLM 74/75 before the plane was repainted at the Front, the tail having been left, however.

An Fw 190A-5/U-3 of 10 (Jabo)./JG 2 at Caen-Carpiquet, France. This aircraft, equipped here with a SD 250 bomb, participated along with 10 (Jabo)/JG 26, (a sister and rival unit) in the raid on Canterbury. This raid took place on October 31st 1942 at the end of the afternoon and the Fw 190s were escorted by 30 fighters from JG 2 and JG 26. There were substantial civilian casualties. The camouflage is RLM 74/75/76 with a green-black spinner in RLM 70 and yellow tactical markings.

76

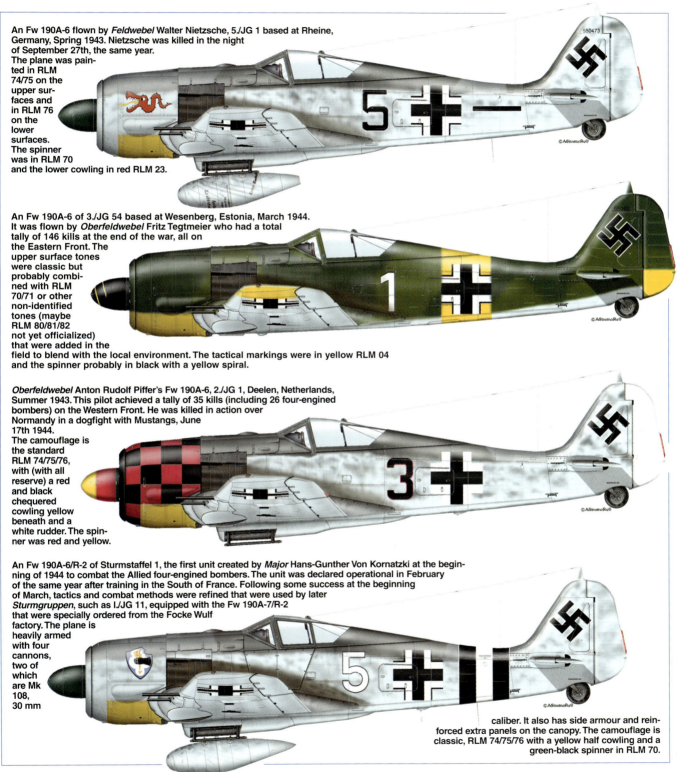

An Fw 190A-6 flown by *Feldwebel* Walter Nietzsche, 5./JG 1 based at Rheine, Germany, Spring 1943. Nietzsche was killed in the night of September 27th, the same year. The plane was painted in RLM 74/75 on the upper surfaces and in RLM 76 on the lower surfaces. The spinner was in RLM 70 and the lower cowling in red RLM 23.

An Fw 190A-6 of 3./JG 54 based at Wesenberg, Estonia, March 1944. It was flown by *Oberfeldwebel* Fritz Tegtmeier who had a total tally of 146 kills at the end of the war, all on the Eastern Front. The upper surface tones were classic but probably combined with RLM 70/71 or other non-identified tones (maybe RLM 80/81/82 not yet officialized) that were added in the field to blend with the local environment. The tactical markings were in yellow RLM 04 and the spinner probably in black with a yellow spiral.

Oberfeldwebel Anton Rudolf Piffer's Fw 190A-6, 2./JG 1, Deelen, Netherlands, Summer 1943. This pilot achieved a tally of 35 kills (including 26 four-engined bombers) on the Western Front. He was killed in action over Normandy in a dogfight with Mustangs, June 17th 1944. The camouflage is the standard RLM 74/75/76, with (with all reserve) a red and black chequered cowling yellow beneath and a white rudder. The spinner was red and yellow.

An Fw 190A-6/R-2 of Sturmstaffel 1, the first unit created by *Major* Hans-Gunther Von Kornatzki at the beginning of 1944 to combat the Allied four-engined bombers. The unit was declared operational in February of the same year after training in the South of France. Following some success at the beginning of March, tactics and combat methods were refined that were used by later *Sturmgruppen*, such as I./JG 11, equipped with the Fw 190A-7/R-2 that were specially ordered from the Focke Wulf factory. The plane is heavily armed with four cannons, two of which are Mk 108, 30 mm caliber. It also has side armour and reinforced extra panels on the canopy. The camouflage is classic, RLM 74/75/76 with a yellow half cowling and a green-black spinner in RLM 70.

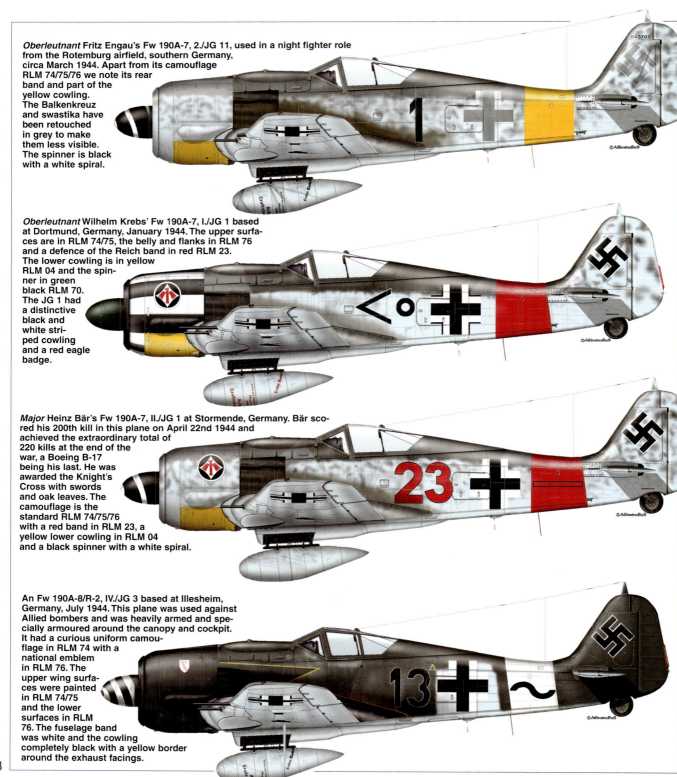

Oberleutnant Fritz Engau's Fw 190A-7, 2./JG 11, used in a night fighter role from the Rotemburg airfield, southern Germany, circa March 1944. Apart from its camouflage RLM 74/75/76 we note its rear band and part of the yellow cowling. The Balkenkreuz and swastika have been retouched in grey to make them less visible. The spinner is black with a white spiral.

Oberleutnant Wilhelm Krebs' Fw 190A-7, I./JG 1 based at Dortmund, Germany, January 1944. The upper surfaces are in RLM 74/75, the belly and flanks in RLM 76 and a defence of the Reich band in red RLM 23. The lower cowling is in yellow RLM 04 and the spinner in green black RLM 70. The JG 1 had a distinctive black and white striped cowling and a red eagle badge.

Major Heinz Bär's Fw 190A-7, II./JG 1 at Stormende, Germany. Bär scored his 200th kill in this plane on April 22nd 1944 and achieved the extraordinary total of 220 kills at the end of the war, a Boeing B-17 being his last. He was awarded the Knight's Cross with swords and oak leaves. The camouflage is the standard RLM 74/75/76 with a red band in RLM 23, a yellow lower cowling in RLM 04 and a black spinner with a white spiral.

An Fw 190A-8/R-2, IV./JG 3 based at Illesheim, Germany, July 1944. This plane was used against Allied bombers and was heavily armed and specially armoured around the canopy and cockpit. It had a curious uniform camouflage in RLM 74 with a national emblem in RLM 76. The upper wing surfaces were painted in RLM 74/75 and the lower surfaces in RLM 76. The fuselage band was white and the cowling completely black with a yellow border around the exhaust facings.

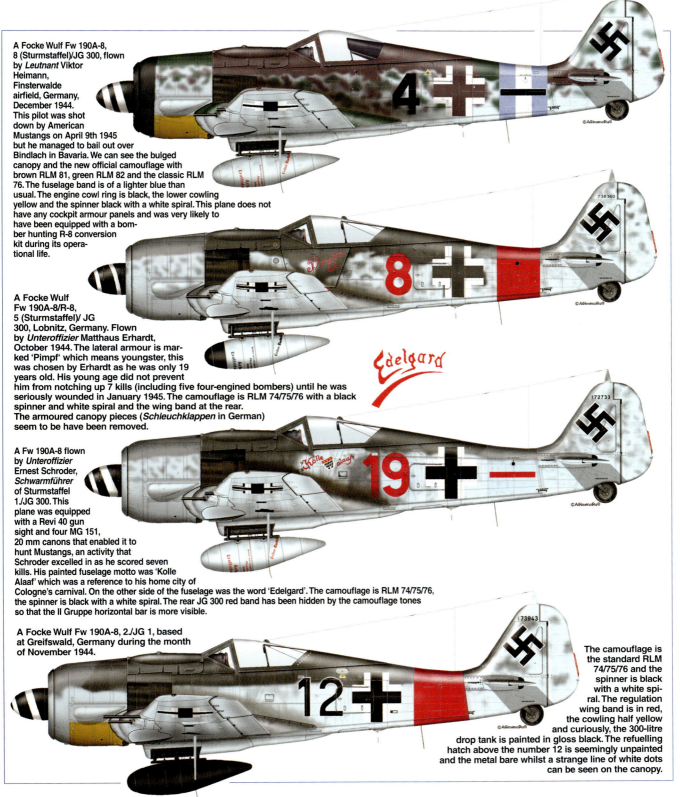

A Focke Wulf Fw 190A-8, 8 (Sturmstaffel)/JG 300, flown by *Leutnant* Viktor Heimann, Finsterwalde airfield, Germany, December 1944. This pilot was shot down by American Mustangs on April 9th 1945 but he managed to bail out over Bindlach in Bavaria. We can see the bulged canopy and the new official camouflage with brown RLM 81, green RLM 82 and the classic RLM 76. The fuselage band is of a lighter blue than usual. The engine cowl ring is black, the lower cowling yellow and the spinner black with a white spiral. This plane does not have any cockpit armour panels and was very likely to have been equipped with a bomber hunting R-8 conversion kit during its operational life.

A Focke Wulf Fw 190A-8/R-8, 5 (Sturmstaffel)/ JG 300, Lobnitz, Germany. Flown by *Unteroffizier* Matthaus Erhardt, October 1944. The lateral armour is marked 'Pimpf' which means youngster, this was chosen by Erhardt as he was only 19 years old. His young age did not prevent him from notching up 7 kills (including five four-engined bombers) until he was seriously wounded in January 1945. The camouflage is RLM 74/75/76 with a black spinner and white spiral and the wing band at the rear. The armoured canopy pieces (*Schleuchklappen* in German) seem to be have been removed.

A Fw 190A-8 flown by *Unteroffizier* Ernest Schroder, *Schwarmführer* of Sturmstaffel 1./JG 300. This plane was equipped with a Revi 40 gun sight and four MG 151, 20 mm canons that enabled it to hunt Mustangs, an activity that Schroder excelled in as he scored seven kills. His painted fuselage motto was 'Kolle Alaaf' which was a reference to his home city of Cologne's carnival. On the other side of the fuselage was the word 'Edelgard'. The camouflage is RLM 74/75/76, the spinner is black with a white spiral. The rear JG 300 red band has been hidden by the camouflage tones so that the II Gruppe horizontal bar is more visible.

A Focke Wulf Fw 190A-8, 2./JG 1, based at Greifswald, Germany during the month of November 1944. The camouflage is the standard RLM 74/75/76 and the spinner is black with a white spiral. The regulation wing band is in red, the cowling half yellow and curiously, the 300-litre drop tank is painted in gloss black. The refuelling hatch above the number 12 is seemingly unpainted and the metal bare whilst a strange line of white dots can be seen on the canopy.

79

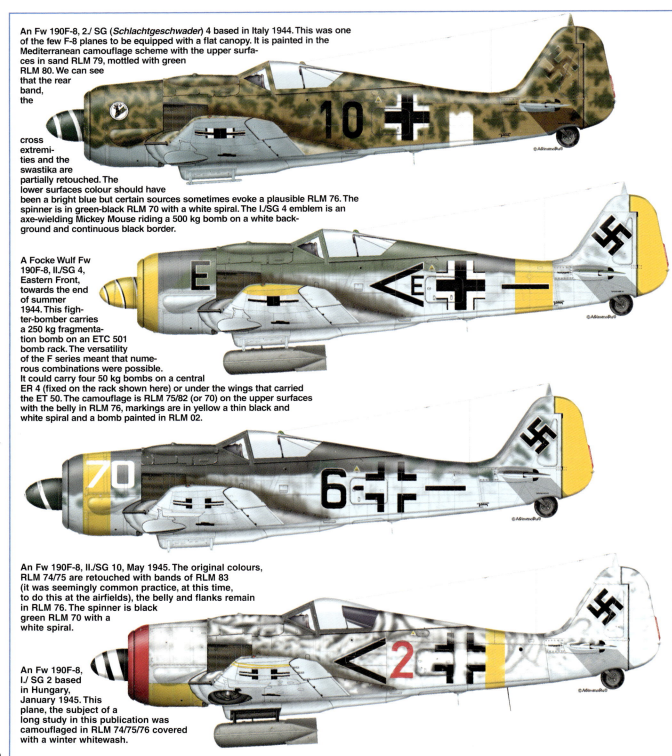

An Fw 190F-8, 2./ SG (*Schlachtgeschwader*) 4 based in Italy 1944. This was one of the few F-8 planes to be equipped with a flat canopy. It is painted in the Mediterranean camouflage scheme with the upper surfaces in sand RLM 79, mottled with green RLM 80. We can see that the rear band, the cross extremities and the swastika are partially retouched. The lower surfaces colour should have been a bright blue but certain sources sometimes evoke a plausible RLM 76. The spinner is in green-black RLM 70 with a white spiral. The I./SG 4 emblem is an axe-wielding Mickey Mouse riding a 500 kg bomb on a white background and continuous black border.

A Focke Wulf Fw 190F-8, II./SG 4, Eastern Front, towards the end of summer 1944. This fighter-bomber carries a 250 kg fragmentation bomb on an ETC 501 bomb rack. The versatility of the F series meant that numerous combinations were possible. It could carry four 50 kg bombs on a central ER 4 (fixed on the rack shown here) or under the wings that carried the ET 50. The camouflage is RLM 75/82 (or 70) on the upper surfaces with the belly in RLM 76, markings are in yellow a thin black and white spiral and a bomb painted in RLM 02.

An Fw 190F-8, II./SG 10, May 1945. The original colours, RLM 74/75 are retouched with bands of RLM 83 (it was seemingly common practice, at this time, to do this at the airfields), the belly and flanks remain in RLM 76. The spinner is black green RLM 70 with a white spiral.

An Fw 190F-8, I./ SG 2 based in Hungary, January 1945. This plane, the subject of a long study in this publication was camouflaged in RLM 74/75/76 covered with a winter whitewash.

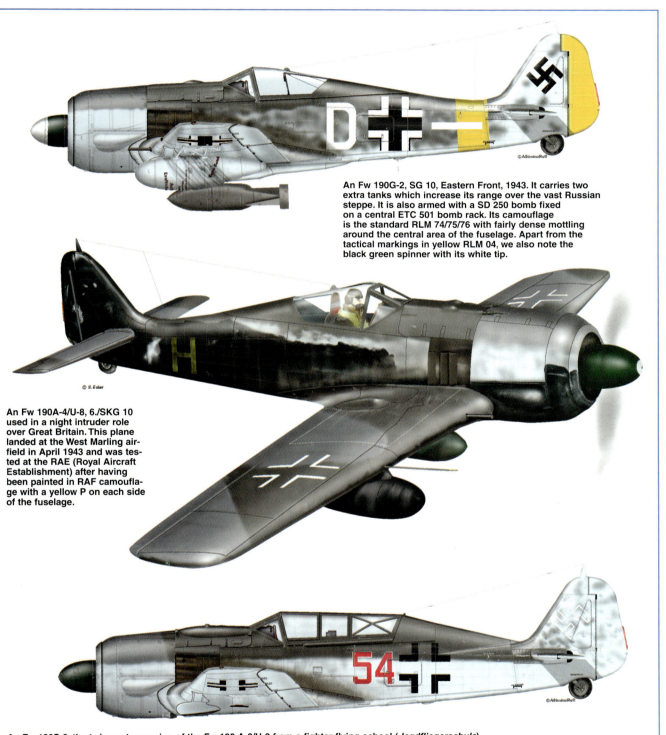

An Fw 190G-2, SG 10, Eastern Front, 1943. It carries two extra tanks which increase its range over the vast Russian steppe. It is also armed with a SD 250 bomb fixed on a central ETC 501 bomb rack. Its camouflage is the standard RLM 74/75/76 with fairly dense mottling around the central area of the fuselage. Apart from the tactical markings in yellow RLM 04, we also note the black green spinner with its white tip.

An Fw 190A-4/U-8, 6./SKG 10 used in a night intruder role over Great Britain. This plane landed at the West Marling airfield in April 1943 and was tested at the RAE (Royal Aircraft Establishment) after having been painted in RAF camouflage with a yellow P on each side of the fuselage.

An Fw 190S-8, the twin seater version of the Fw 190 A-8/U-8 from a fighter flying school (*Jagdfliegerschule*) end of 1944. The camouflage is RLM 74/75/76 with a black green spinner in RLM 70.

BIBLIOGRAPHY

● **MONOGRAPHS ON THE AU FW 190**
— **Focke-Wulf Fw 190A/F,** *Aero Details n° 6*
— **Focke Wulf Fw 190A/F/G,** *Model Art n° 316*
— **Focke Wulf Fw 190,** *Squadron Signal in Action n° 19*
— **Focke Wulf Fw 190,** *Squadron Signal in Action n° 170*
— **Focke-Wulf Fw 190A/F,** *Squadron Signal, Walk Around n° 22*
— **Focke-Wulf Fw 190,** *Waffen Arsenal n° 95*
— **Focke-Wulf Fw 190,** *Burindo n° 6 et 78*
— **Focke Wulf Fw 190A/F/G,** *Monografie Lotnicze n° 17* (1re Pt.)
— **Focke Wulf Fw 190A/F/G,** *Monografie Lotnicze n° 18* (2e pt.)
— **Focke-Wulf Fw 190F,** *Monogram Close-Up n° 8*
— **Focke-Wulf 190,** *Monogram Close-Up n° 18*
— **Focke Wulf Fw 190G,** *JaPo*
— **Focke Wulf Fw 190A,** *Modelpres n° 7*
— **Focke Wulf Fw 190A,** Frank Osman, *Archive*
— **Focke Wulf Fw 190 1943-1945,**
Morten Jessen, *Luftwaffe At War n° 13*
— **Focke-Wulf Fw 190 Aces of the Russian Front,**
John Weal, *Osprey Aircraft of the Aces n° 6*
— **Focke-Wulf Fw 190 Aces of the Western Front,**
John Weal, *Osprey Aircraft of the Aces n° 9*

● **GENERAL PUBLICATIONS ON THE FW 190**
— **Sturmstaffel 1 Reich Defence 1943-1944.**
The War Diary, *Air War Classics*
— **Defenders of the Reich Jagdgeschwader 1, 1939-1942** (Vol. 1),
Air War Classics
— **Defenders of the Reich Jagdgeschwader 1, 1943** (Vol. 2),
Air War Classics
— **JG 2 Richthofen 1942-1943,** Krzysztof Janowicz, *Kagero n° 4*

— **Encyclopédie Illustrée de l'Aviation** (Vol. 3, page 526)
— **Die Deutsche Tagjagd,** Werner Held, *Motor buch Verlag*
— **Die Deutsche Nachtjagd,** Werner Held, *Motor buch Verlag*
— **German Aircraft Landing Gear,** Günter Sengfelder, *Schiffer*
— **German Aircraf Interiors 1935-1945** (Vol. I),
Keneth Merrick, *Monogram*
— **Warplanes of the Third Reich,** William Green, *MacDonald & Janes*
— **Planes of the Luftwaffe Fighter Aces** (Vol. I et II),
Bernd Barbas, *Kookaburra*

● **LUFTWAFFE CAMOUFLAGE & MARKINGS**
— **The Official Monogram Painting Guide to German Aircraft 1935-1945,** Merrick & Hitchcok, *Monogram*
— **Luftwaffe Color Chart Official.** Eagle Editions Ltd.
— **Luftwaffe Colors 1935-1945,** Michael Ullman, *Hikoki*
— **Luftwaffe Camouflage & Markings 1935-1945** (Vol. I/II/III),
Smith & Gallapsy, *Kookaburra*
— **Camouflage & Markings Luftwaffe 1935-1945** (Vol. I/II/III/IV/V),
AJ Press
— **Luftwaffe Codes, Markings & Units 1939-1945,**
Barry C. Rosch, *Schiffer*
— **Luftwaffe Rudder Markings & Units 1936-1945,**
K. Ries & E. Obermaier, *Schiffer*
— **Luftwaffe Aircraft Fighters in Profile,**
C. Sundin & Ch. Bergström, *Schiffer*
— **More Luftwaffe Aircraft Fighters in Profile,**
C. Sundin & Ch. Bergström, *Schiffer*

● **OTHER REFERENCES TO BE CONSULTED**
— **Colour chart Federal Standard Colors** *595a.*

Supervision and adaptation Anis EL BIED - Translation Marie-Françoise VINTHIERE and Lawrence BROWN
Design, creation and production Jean-Marie MONGIN,
© Histoire & Collections 2006

All rights reserved. No part of this publication can be transmitted or reproduced without the written consent of the Author and the Publisher.
ISBN: 2-915239-58-4
Publisher's number: 2-915239

A book edited by
HISTOIRE & COLLECTIONS
SA au capital de 182 938, 82 €
5, avenue de la République F-75541 Paris Cédex 11
N° Indigo 0 820 888 911
0,118 € TTC / MN
Fax 01 47 00 51 11
www.histoireetcollections.fr

This book has been designed, typed, laid-out and processed by *Histoire & Collections,* fully on integrated computer equipment.

Printed by Zure, Spain, European Union.
April 2006